石油石化企业现场安全督导系列丛书

安全督导推荐做法及常用安全模型

Recommended Leadership Site Visit Program and Safety Models

赵宏展　贺晓珍　杨意峰　编著

中国石化出版社

内 容 提 要

本书为《石油石化企业现场安全督导系列丛书》之一，在介绍基于“安全交谈”的“管理人员现场安全督导方案”及其具体做法和注意事项的基础上，对管理巡查、管理层检查和管理审核的要点进行了描述，重点阐述了做好管理人员现场安全督导需要了解的“危害控制层序”“危险轨迹”“领结图分析法”“屏障”“保护层分析”“安全例证”等6个常用安全模型。

本书内容为中英文对照，便于读者将中文和英文结合在一起阅读，以便准确理解其内容。

本书可用于石油石化企业各级管理人员自学或培训，也可作为其他行业的管理人员以及石油石化大专院校工科专业及其他院校安全相关专业师生的安全管理类参考书。

图书在版编目（CIP）数据

安全督导推荐做法及常用安全模型 / 赵宏展，贺晓珍，杨意峰编著 .—北京：中国石化出版社，2019. 8
（石油石化企业现场安全督导系列丛书）
ISBN 978-7-5114-5447-8

Ⅰ. ①安… Ⅱ. ①赵… ②贺… ③杨… Ⅲ. ①石油化工-安全管理 Ⅳ. ①TE687

中国版本图书馆 CIP 数据核字（2019）第 170229 号

中国石化出版社出版发行

地址：北京市东城区安定门外 58 号
邮编：100011　电话：(010)57512500
发行部电话：(010)57512575
http://www.sinopec-press.com
E-mail:press@sinopec.com
北京科信印刷有限公司印刷
全国各地新华书店经销

*

850×1168 毫米 32 开本 6.375 印张 141 千字
2019 年 11 月第 1 版　2019 年 11 月第 1 次印刷
定价：36.00 元

前言
Preface

企业健康安全环境(HSE)管理工作的属地管理和直线责任原则已经深入人心，很多企业也高度重视“有感领导”(指有安全认知的领导)，不断强化各级管理人员对HSE工作的示范作用。越来越多企业的管理人员经常深入生产作业现场，实地调研和指导HSE管理工作，践行“有感领导”。随着企业HSE管理体系越来越成熟，不同角度、不同深度、不同方式的各级管理人员开展的HSE检查和HSE审核也越来越丰富多样，对于生产作业场所的HSE风险管控起到了至关重要的作用。

上述背景产生了两个问题：一是石油和石化企业各级管理人员在企业HSE管理体系框架下如何更精准有效地发力？二是石油和石化企业有哪些共性的HSE知识模块值得总结和提炼？

编著者基于10余年的海外项目工作经验，在广泛调研国内外大量文献资料的基础上，通过跟诸多资深项目管理人员和资深HSE专业人员的沟通交流，编写了本丛书。本丛书呈上了对上述两个问题的初步回应，推荐了基于“安全交谈(Safety Conversation)”的“管理人员现场安全督导方案(Leadership Site Visit Program)”，包括：具体做法、注意事项、支持性信息(知识模块)。

在安全氛围(Safety Climate)的三“心”模型[1]基础上，我们坚信：基于“安全交谈”的管理人员现场安全督导，可以通过管理人员对于安全问题的“关心”，激励员工对于安全问题的“上心”，并最终营造企业或项目对于安全问题的“上下一心”。

人们期待“安全保障”，并提出了“安全保障”的需求和要求。实际上，只有“共建安全”才能获得更加牢靠的“安全保障”，因为“安全保障”的需求者和供给者是不可分离的。在属地管理、直线责任、“有感领导”的基础上，人们还必须坚持“安全共建”和“安全共享”的原则，即每个人都坚信“我需要安全，安全需要我”。

本丛书呈上的中英文对照的支持性信息(知识模块)共分为八组，分别是：

(1) HSE 风险管理工具和过程；

(2) HSE 风险贡献因素和升级因素；

(3) 旅程管理和道路交通安全；

(4) 人员健康和保护；

(5) 工作场所常见危害；

(6) 潜在高后果活动；

(7) 危险有害物质；

(8) 火灾爆炸保护。

---

[1]于广涛，李永娟．安全氛围三“心”模型的构建与检验[J]．中国安全科学学报，2009，19(9)：28.

本丛书“臻选”的支持性信息(知识模块)融入了以下内容:(危害)控制层序(Hierarchy of Controls)的原则、设计保障安全(Prevention through Design)的做法、领结图(Bowtie Diagram)的精髓、保护层分析法(Layer of Protection Analysis)的思路、屏障(Barrier)的要义。此外,安全例证(Safety Case)不仅仅是英国等发达国家法律规定的具体做法,更是一种风险管理思路。危险轨迹(Line of Fire)既是生产作业场所的一种“常态”,也是很多人身伤害的本质,目前已经作为国际油气生产商协会(IOGP)在2018年发布的九条保命法则(Life-Saving Rules)之一而更加受到重视。管理巡查(Management Walkthrough)、管理层检查(Management Inspection)和管理审核(Management Audit),三者密切相关,也经常被混淆,本丛书对其区别和联系也做出了解读。

管理人员现场安全督导推荐做法、常用安全模型以及八组支持性信息(知识模块)等内容顺序编排在本丛书的相应分册当中。四个分册的名称具体如下:

- 《安全督导推荐做法及常用安全模型》
- 《基本HSE风险因素及管理》
- 《工作场所常见危害及潜在高后果活动》
- 《危险有害物质及火灾爆炸保护》

本丛书的英文内容参考了以下政府组织和行业协会公开出版的英文文献资料,在此表示诚挚感谢!

◇ 英国健康安全执行局(UK HSE，UK Health and Safety Executive)

◇(美国)职业安全与卫生管理局(OSHA，Occupational Safety and Health Administration)

◇ 美国化学工程师协会(AIChE，American Institute of Chemical Engineers)

◇ 国际油气生产商协会(IOGP，International Association of Oil & Gas Producers)

◇ 国际钻井承包商协会(IADC，International Association of Drilling Contractors)

本丛书的写作还参考了其他大量国内外文献资料，在此对原著者也深表感谢!

回望过去，本丛书英文内容的最终定稿是一个漫长的积淀和筛选过程，历时10余年时间。2010年1月，编著者之一刚刚取得安全工程博士学位，作为唯一的中方人员，有幸被指派组建一支国际化的HSE团队，服务于一个员工来自30多个国家的中东地区油田开发生产项目。先后有来自英国、美国、法国、加拿大、澳大利亚、新西兰等国家的60多位资深HSE专业人士以油田员工或知名咨询公司顾问的身份，成为这支国际化HSE团队的一员。2012年12月和2013年6月，本丛书的另两位编著者先后加入这支HSE团队，成为协调和领导这支国际化HSE团队的核心力量，但英文仍然是覆盖整个HSE团队的唯一工作语言。在开放、包容、交流、共享、相互尊重的基础上，这支 HSE 团队整体所能提供的 HSE 支持和服务实现了最大程度的国际接轨。总之，本丛书可以说是编著者经过10余年开放融通的海外项目磨砺后交上的一份答卷。

本丛书“臻选”的支持性信息(知识模块)采用中英文对照的形式编写和排版，目的是希望读者能够将中文和英文结合在一起阅读，提取其中的“最大公约数”，从而更加精准地理解本丛书所提供信息和知识的本意，更好地辅助和支持管理人员开展现场安全督导。

本丛书最终得以公开出版得到了众多资深管理人员、中外专家和 HSE 同仁的支持和参与，凝聚了众人的智慧和心血，编著者不过是在知识的沙滩上捡拾贝壳的三人，我们虔诚地把认为有价值的知识贝壳收集起来，经过筛选和加工以中英文双语的形式呈现给广大读者。感谢默默无闻支持编著者完成本丛书的所有人！

本丛书主要适用于石油石化企业各级管理人员自学或培训之用，也可作为其他行业的管理人员以及石油石化院校工科专业及其他院校安全相关专业师生的安全管理类参考书。

管理人员现场安全督导工作有着很强的知识性和实践性，书中难免存在疏漏之处，敬请批评指正。如有任何意见和建议，欢迎联系以下电子邮件地址：safetyview@ 163. com。

目 录

5 屏障

6 保护层分析(LOPA)

7 安全例证

附录

Contents

5 BARRIER

6 LAYER OF PROTECTION ANALYSIS (LOPA)

1

INTRODUCTION TO LEADERSHIP SITE VISIT PROGRAM (LSVP)

管理人员现场安全督导工作（LSVP）介绍

1.1 Brief

Leadership Site Visits (LSVs), as part of the Leadership Site Visit Program (LSVP), are intended to bridge the gap between the more formal Management Inspection system and the informal Safety Observation system which are essential parts of the leadership and management process.

LSVs are key components of the HSE-MS assurance process in which managers physically verify that operations are conducted in accordance with the HSE-MS policy, standards and procedures. The results of the LSVs are formally documented and effectively followed-up through the LSVP Tracking System.

1.1.1 Purpose

The Leadership Site Visit Program (LSVP) is intended to help Senior Managers and Middle Managers go out on site more often and to enhance the interaction between all Managers (both Senior and Middle Managers) and the workforce. The main purpose of the program is to:

- Enable all Managers to become more involved in day-to-day operations at the facilities.
- Ensure all Managers go to site on a regular basis.
- Allow all Managers to visibly promote HSE matters while on site.

1.1 概要

作为管理人员[1]现场安全督导工作(LSVP)的一部分，管理人员现场安全督导(LSVs)的意图是在非常正式的管理人员检查方法和非正式的安全观察方法之间架起桥梁。它们都是领导和管理过程的必要组成部分。

管理人员现场安全督导(LSVs)是HSE管理体系(HSE-MS)保证过程的关键组成部分。在开展管理人员现场安全督导(LSVs)的过程中，管理人员亲身核实作业是否遵照HSE管理体系(HSE-MS)政策、标准和程序的要求执行。管理人员现场安全督导(LSVs)的结果要正式记录下来，并通过管理人员现场安全督导工作(LSVP)追踪系统进行有效跟踪。

1.1.1 目的

管理人员现场安全督导方案[2](LSVP)的意图是帮助高层管理人员和中层管理人员更频繁地深入作业现场，强化不同层级管理人员(高层管理人员和中层管理人员)和作业现场工作人员之间的互动。其主要目的如下：

- 为管理人员更多地深入生产作业场所的日常运行提供措施方法。
- 确保管理人员经常性深入生产作业现场开展工作。
- 为管理人员在生产作业现场显著推进HSE事项(的管控)提供条件。

---

[1]主要指的是非常驻生产作业场所但对生产作业场所负有监管责任的各级管理人员，通常包括高层管理人员和中层管理人员等。

[2]根据上下文，LSVP翻译为“管理人员现场安全督导方案”或“管理人员现场安全督导工作”。

- Promote interaction between all Managers and workers on HSE matters.
- Create awareness of general attitudes and concerns of the workforce on HSE matters.

LSVs can enhance HSE teamwork, workforce culture and willingness to change and improve.

1.1.2 Safety Topics

This book describes the requirements to be applied for conducting LSVs. LSVs Should be conducted by all relevant Managers.

There are 33 selected Safety Topics covered by the LSVP. A LSV Prompt Card sample (Appendix 1) is provided that can be used as the basis for conducting the site visits.

The 33 Safety Topics fall into eight broad groups. The Safety Topics in Group 1 are those which should be commonly understood by all participants. The subsequent groups include topics which require an increasing level of knowledge and may require a certain level of technical competence to understand the topic. Participants performing LSVs for these topics must ensure they fully understand the topic in order to make best use of the visit.

(1) Group 1: HSE Risk Management Tools and Processes

Workplace HSE risk management tools and processes are commonly used at site to manage risks associated with routine, non-routine and emergency situations.

- 推进管理人员和生产作业现场工作人员针对 HSE 事项的良性互动。
- 使工作人员意识到对于 HSE 事项的应有态度和关切。

管理人员现场安全督导(LSVs)可以增强 HSE 方面的团队合作、工作人员(安全)文化素养以及改变和提升的意愿。

1.1.2 安全专题

本书描述了开展管理人员现场安全督导(LSVs)的要求。相关管理人员均应开展现场安全督导(LSVs)。

本书介绍了管理人员现场安全督导工作(LSVP)可能涉及的 33 个安全专题，提供了管理人员现场安全督导提示卡的样例(附录 1)。管理人员现场安全督导提示卡应作为开展管理人员现场安全督导工作的基础。

本丛书的 33 个安全专题划分为 8 个组别。管理人员均应熟知组别 1 的安全专题。随后组别的安全专题涉及的知识广度和深度不断增加，有些安全专题可能还需要特别的技术背景辅助理解。为了高质量地完成管理人员现场安全督导(LSVs)，管理人员必须完整全面的理解相关安全专题。

(1) 组别 1：HSE 风险管理工具和过程

为管控常规、非常规以及紧急事态相关的风险，工作场所 HSE 风险管理工具和过程得到广泛应用。

The types of work controlled in the workplace can be categorised as routine or non-routine. Routine activities are frequently performed on a day to day basis that broadly cover the entire spectrum of risks. Non-routine activities are infrequent, often associated with maintenance activities and considered to be of high risk potential. Non-routine activities are controlled in the Permit to Work (PTW) system and require a Job Hazard Analysis (JHA), a form of risk assessment, to be conducted as part of the permit authorisation. Routine work activities may still be subject to a JHA but do not require a PTW.

Toolbox Talks (TBT), a type of workplace meeting, are required prior to commencing routine and non-routine work activities. The aim of the TBT is to discuss the work ahead and to discuss the risk controls that will be in place to control the work. HSE Briefings are normally incorporated into the Daily Operations Meeting to provide a means for discussing HSE programmes and risks affecting the site.

During work, incidents can arise which unless controlled have the potential to become emergency situations requiring assistance from others at site. In some cases, the emergency is so severe that it becomes a crisis situation that overwhelms the sites' response capability requiring external assistance (i.e. ambulance, fire services, Medevac, etc.). Risk assessment can provide focus, direction and guidance for planning a response to site emergency and crisis situations.

The following Safety Topics fall into Group 1:

01 Job Hazard Analysis

02 Permit to Work

03 Workplace HSE Meetings

04 Emergency Management and Crisis Management

工作场所受控的工作类型可以分为常规和非常规两大类。常规活动是指经常性地每天都在进行的活动，这些活动大致覆盖全部风险类别。非常规活动不经常发生，经常跟维修活动相关，被认为具有潜在高风险。非常规活动通过作业许可(PTW)制度加以控制，需要完成基本的风险评估——工作危害分析(JHA)，作为批准作业许可的必要步骤。常规活动可能只需要完成工作危害分析(JHA)，但并不需要作业许可(PTW)。

工具箱会议(TBT)是工作场所会议类型之一，需要在开展常规和非常规工作活动之前进行。工具箱会议(TBT)的目的是讨论即将开展的工作，以及讨论控制拟开展工作所需到位的风险管控措施。通常，HSE 择要说明会跟每日作业例会整合在一起，用以讨论作业 HSE 方案及影响现场的风险。

工作期间可能发生某些事件，如不加以控制，就可能演变为紧急事态。这些事件需要现场其他人员的援助。在某些情况下，紧急事态非常严重，以至于演变为危机事态，超出了现场的响应能力，需要外部援助(比如，救护车、灭火服务、医疗转运等)。为做好现场紧急事态和危机事态的响应计划，风险评估可以提供聚焦点、方向和指南。

以下安全专题归入组别 1：

01 工作危害分析

02 作业许可

03 工作场所 HSE 会议

04 应急管理和危机管理

(2) Group 2: Contribution Factors and Escalation Factors for HSE Risk

Workplace and environmental contributing factors with potential for degradation or failure of HSE risk controls can lead to a reduced level of safety performance.

Under normal conditions risk controls behave in a bounded and predictable way. Changes in workplace and environmental conditions, including Unfit Workers, Adverse Weather, Night Time Working, Working Alone, etc. can contribute to the rapid degradation of normally robust risk controls leading to their failure and escalation of the risk. To prevent escalation it is important to have processes in place to build additional defences to these risk controls prior to the onset of workplace and environmental changes.

The following Safety Topics fall into Group 2:

05 Fitness for Task

06 Adverse Weather

07 Night Working

08 Working Alone

09 Management of Change

(3) Group 3: Journey Management and Land Transport Safety

Land transport operations and vehicle journeys performed on oilfield roads and public highways, pose high safety risks that must be effectively managed.

Land transport operations involving light vehicles and heavy vehicles that carry people, loads or dangerous substances have traditionally posed a high safety risk resulting in numerous fatalities. The primary focus of land transport safety is concerned with vehicle safety specifications, driver competency, road and weather conditions.

(2) 组别 2：HSE 风险贡献因素和升级因素

工作场所自身和环境方面的贡献因素可能导致 HSE 风险管控措施降级[1]或失效，引起安全绩效水平降低。

在正常条件下，风险管控措施以受约束及可预测的方式发挥作用。工作场所自身和环境的变化，包括工作人员身体条件不适合岗位要求、恶劣天气、夜间作业、独自作业等，可能引起本来稳健的风险管控措施迅速降级、失效，造成风险升级。在工作场所自身和环境开始变化之前，为预防风险升级，在通常风险管控措施的基础上，建立额外防范措施并确保其落实到位是至关重要的。

以下安全专题归入组别 2：

05 健康适岗

06 恶劣天气

07 夜间作业

08 独自作业

09 变更管理

(3) 组别 3：旅程管理和道路交通安全

在油田内部道路和公共道路上进行的运输作业和道路旅程，具有高的安全风险，必须有效管控。

道路运输作业包含轻型车辆和重型车辆，装载人员、货物或危险物质，在安全方面通常存在导致大量死亡的高风险。道路运输安全的主要关注点在于车辆安全规格、驾驶人员能力、道路和天气条件。

[1] 也翻译为退化。

For non-routine, high risk potential, long road journeys then journey management is an effective means of managing the risk of a journey. Journey management involves selecting the lowest risk route (of alternatives), risk assessing the route and putting in place risk controls along the way. The journey is managed and monitored by a Journey Manager who maintains contact with the vehicle and driver at key points along the journey. Failure to make contact initiates an emergency response and recovery.

The following Safety Topics fall into Group 3:

10 Journey Management

11 Land Transport Safety

(4) Group 4: Personal Health and Protection

Personal health is a high priority for all workers in the workplace. The exposure of personnel to physical, chemical and biological hazards without protection can harm their health either suddenly, or more gradually over the longer term.

Personal health may be impacted through exposure to work place physical hazards. In the case of heat stress, the hazard is experienced over a short duration which if uncontrolled could, in the severest case, leads to worker fatality. In the case of noise, the hazard is experienced over a period of years before the irreversible effects of noise induced hearing loss may be experienced.

Noise levels from equipment should be considered during design to minimise noise levels during operation, however, this is not always possible and as such Personal Protective Equipment (PPE) may be required to provide an adequate level of protection. Whilst mandatory to wear in designated high noise level areas, PPE should always be considered as the last resort for hearing protection.

对非常规、潜在高风险、远距离的旅程，旅程管理提供了有效的管理旅程风险的方式，包括在替代路线中选择最低风险的路线、路线风险评估、沿路落实风险管控措施。例如：旅程监控人员跟车辆和驾驶人员在旅途的关键点保持联系。联系失败需启动应急响应和恢复措施。

以下安全专题归入组别3：

10 旅程管理

11 道路交通安全

(4) 组别4：人员健康和保护

人员健康是工作场所人员的头等大事。人们在没有得到妥善保护的情况下，暴露于物理、化学和生物危害中，其健康会以突然或长期渐进的方式受到伤害。

人员健康可能会由于暴露于工作场所的物理危害而受到影响。就热应激来说，该危害即便是在较短的时间内未受控制，在最严重的情况下也可能导致死亡。就噪声来说，在产生不可逆转的噪声引致的听力损失之前，患者可能已经数年暴露于噪声危害。

设备的噪声等级应该在设计阶段考虑将其最小化，然而实际情况并非如此理想，因此需要个体防护装备(PPE)提供足够的保护。虽然在指定的高噪声等级区域需要强制佩戴个体防护装备(PPE)，但是个体防护装备(PPE)始终是听力保护的最后防线。

Food safety generally requires others who prepare food to put in place protective measures to ensure that food poisoning, or other food related sickness, are avoided.

The following Safety Topics fall into Group 4:

12 Food Safety

13 Heat Stress

14 Noise and Hearing Conservation

15 Personal Protective Equipment

Safety topics in Groups 1-4 are discussed in Book: Fundamental HSE Risk Factors and Management.

(5) Group 5: Common Hazards at Workplaces

The workplace contains many common hazards that are frequently encountered. Even though personnel may be exposed to these common hazards on a daily basis, it doesn't mean that any less attention should be provided in managing their risks.

Common workplace hazards whilst routinely encountered have varying levels of HSE risk. Despite being common they should still be subject to JHA with preventative and mitigative controls put in place to ensure a safe method of work. When associated with non-routine work activities, risks from the common workplace hazards should be considered in the JHA as part of the PTW system.

The following Safety Topics fall into Group 5:

16 Slips Trips and Falls

17 Dropped Object Prevention

18 Electrical Safety

19 Portable Power Tools

20 Work Equipment Safety

食品安全通常需要准备食物的人员落实保护他人健康的措施，避免食物中毒或其他食物相关疾病的发生。

以下安全专题归入组别4：

12 食品安全

13 热应激

14 噪声和听力保护

15 个体防护装备

组别1~4的安全专题包含在丛书分册《基本HSE风险因素及管理》中。

(5) 组别5：工作场所常见危害

工作场所有很多常见危害。虽然现场工作人员每天都暴露于这些常见危害中，但这不意味着这些常见危害的风险需要较少的关注。

工作场所常见危害有着不同的HSE风险水平。尽管很常见，但对它们仍应进行工作危害分析(JHA)；为确保作业安全，危害预防和削减措施应落实到位。当常见危害跟非常规活动相关时，它们的风险防控措施除了工作危害分析(JHA)，还需要执行作业许可(PTW)制度，前者是后者的一部分。

以下安全专题归入组别5：

16 滑倒、绊倒和摔倒(跌落)

17 落物防范

18 电气安全

19 便携式动力工具

20 工作设备安全

21 Pressure Systems Safety

22 Moving and Energised Equipment

(6) Group 6: Activities with Potential Severe Consequences

Less common hazards that may be infrequently encountered can pose a high injury and multiple fatality risk to personnel. Great focus and attention should be placed on managing non-routine risks through the formal PTW system.

The PTW System is particularly focused on non-routine workplace activities that are often high risk due to their potential for multiple fatalities. All non-routine activities with potential severe consequences are managed by the PTW system. Work at Height, Excavation, Confined Space Entry, Lifting Operations and Isolation all require certificates as part of the PTW system. Each certificate provides details of activity specific risk controls as supported by a JHA. Further details on PTW and JHA risk management processes are provided in Group 1.

The following Safety Topics fall into Group 6:

23 Work at Height

24 Lifting Operations

25 Excavations

26 Isolation

27 Gas Testing

28 Confined Space Entry

Safety topics in Groups 5 and 6 are discussed in Book: Common Hazards and Activities with Potential Severe Consequences in Workplaces.

21 压力系统安全

22 机动设备

（6）组别 6：潜在高后果活动

一些不那么常见的危害可能产生严重伤害和多人死亡，因此非常规危害需要通过正式作业许可（PTW）制度予以高度关切和注意。

作业许可（PTW）制度特别关注工作场所非常规活动。由于这些活动可能导致多人死亡，它们常常是高风险的。所有潜在高后果的非常规活动，均需通过作业许可（PTW）制度加以管控。高处作业、动土作业、进入受限空间、吊装作业和隔离作业，都需要（使用）作为作业许可（PTW）系统一部分的专项作业单。专项作业单提供专项作业活动特定的、基于工作危害分析（JHA）的风险管控措施细节。作业许可（PTW）和工作危害分析（JHA）的更多细节，详见安全专题组别 1。

以下安全专题归入组别 6：

23 高处作业

24 吊装作业

25 动土作业

26 隔离作业

27 气体检测

28 进入受限空间

组别 5 和组别 6 的安全专题包含在丛书分册《工作场所常见危害及潜在高后果活动》中。

(7) Group 7: Hazardous Substances

Hazardous and toxic substances are an integral part of the oil and gas industry, from the Major Accident Hazards (MAHs) associated with toxic and flammable reservoir fluids that are part of the process, to the wide variety of chemical hazards stored or used at the facilities.

Chemicals are widely used in the oil and gas industry often as a means to mitigate HSE risks. Despite their beneficial properties, many chemicals are also hazardous and may cause harm to human health and the environment. Chemical risk assessments ensure that controls are put in place across the chemical life cycle to ensure that chemicals are stored, handled, used and disposed of in a safe and environmentally sound fashion.

Hazardous substances may have flammable and explosive properties. Large quantities of highly flammable substances have the potential to lead to Major Accidents. Group 8 provides further details on Major Accidents from fires and explosions.

The following Safety Topics fall into Group 7:

29 Chemicals Safety

30 Hydrogen Sulphide (Toxic Gas) Safety

(8) Group 8: Fire and Explosion Protection

The processing and storage of flammable and explosive hydrocarbons has the potential to lead to a Major Accident resulting in a major hydrocarbon fire or explosion. Activities associated with combustible materials in offices and buildings have the potential to lead to major building fires that can cause fatalities and extensive property damage.

(7) 组别 7：危险物质

危险和有毒物质是石油天然气行业必不可少的组成部分。这些物质既包括重大事故危害(**MAHs**)，也包括品种繁多的其他化学危害。前者跟有毒和易燃的地下油藏流体有关，它们是生产工艺流程的一部分；后者也存储或应用于各生产设施。

化学品在石油天然气行业被广泛使用，通常被作为削减 HSE 风险的手段。它们虽具备有益特性，但许多化学品也是危险的，可能对人的健康产生伤害或对环境造成破坏。化学品风险评估确保化学品的全生命周期管控措施到位并有效实施，确保化学品以安全和环境友好的方式存储、管理、使用、废弃。

危险物质可能具有易燃易爆特性，大量极易燃的物质有导致重大事故的可能性。关于火灾爆炸重大事故，组别 8 将提供进一步的细节。

以下安全专题归入组别 7：

29 化学品安全

30 硫化氢(有毒气体)安全

(8) 组别 8：火灾和爆炸保护

处理和储存易燃和易爆烃类，有产生着火或爆炸重大事故的潜在风险。在办公室和建筑物中，涉及可燃物料的活动，也有发生建筑物重大火灾的潜在风险，可能导致亡人和大量财产破坏。

Major Accidents are catastrophic events mainly caused by losing control of the hydrocarbon fluids within process containment systems. The release of large quantities of highly flammable and potentially toxic process fluids can lead to major fire, explosion and toxic risks that have the potential to severely impact people, the environment, assets or the company's reputation.

Fuel storage tanks operated at atmospheric temperatures and pressures, contain flammable substances (e.g. diesel) considered to be a fire and explosion hazard. Due to the lower hydrocarbon storage inventory, higher flashpoint and reduced ignition potential, fuel storage tanks are not normally considered to be of Major Accident potential.

Offices and buildings need to be designed and operated to provide an acceptable level of fire safety and to minimise the risks from heat and smoke. The main objective is to reduce the potential for fire related death or injury to the occupants of a building and others who may become involved, such as the fire and rescue service, to acceptable limits.

The following Safety Topics fall into Group 8:

31 Fire Protection——Buildings

32 Fire and Explosion Protection——Fuel Storage Tanks

33 Fire and Explosion Protection——Hydrocarbon Facilities

Safety topics in Groups 7 and 8 are discussed in Book: Hazardous Substances and Fire & Explosion Protection.

1.1.3 Supporting Information

Each safety topic has been provided with supporting information in this set of books. The purpose of the supporting information is to provide briefing and pre-read material for each safety topic.

（石油天然气行业的）重大事故是灾难性的，这些事故主要是由密闭工艺系统中的烃类流体失去控制引起的。大量极易燃和可能有毒的工艺流体意外释放，可能导致重大火灾、爆炸和毒害风险，可能严重影响人员、环境、资产或企业声誉。

常温常压下运行的燃油储罐，包含被认为有火灾爆炸危险的易燃物质（比如：柴油）。由于较低的烃类存储量，更高的闪点和降低的点火可能性，通常不认为其有重大事故的潜在危险。

办公室和建筑物的设计和运行，需要提供可接受的消防安全水平，最小化热量和烟雾产生的风险。以上工作的主要目的是降低对建筑物常驻人员及其他可能涉及人员（比如消防和救援人员）可能造成的火灾相关死亡或伤害（风险）至可接受的限度。

以下安全专题归入组别 8：

31 火灾保护——建筑物

32 火灾和爆炸保护——燃油储罐

33 火灾和爆炸保护——涉及烃类的生产设施

组别 7 和组别 8 的安全专题包含在丛书分册《危险有害物质和火灾爆炸保护》中。

1.1.3 支持性信息

本套丛书为 33 个安全专题均提供了支持性信息。支持性信息的目的是为了介绍清楚每个安全专题的基本信息，为开展管理人员现场安全督导提供预习材料。

1.2 Requirements for the Leadership Site Visit Program (LSVP)

1.2.1 Participating Parties

(1) Senior Management

Senior Management are required to take part in the LSVP. They shall complete a minimum of one visit per quarter, however additional visits can be completed depending upon the requirements of the business (i.e. due to incidents at site, audit findings, etc.).

Senior Managers shall produce a timetable of when the visits will be completed and which locations will be visited. A sample of "Leadership Site Visit Matrix" is included in Table 1-1. The Visit Matrix shall be updated throughout the year to show progress and identify any necessary changes to the plan.

(2) All Departments

All departments shall determine which of the safety topics are relevant to their activities. Department Managers may include additional topics as needed, for instance where the activities of another department directly impact the activities of their department.

Middle Managers shall complete the full LSVP topics for their department during each calendar year. It is not necessary for other participants of the department to complete each Leadership Safety Visit (LSV) topic in a single year.

Where a department has numerous activities for a single safety topic through a single year, the Department Manager may decide to perform multiple LSVs on that topic through the year. It would be advantageous to the LSVP if these visits are completed by different people within the department so that different viewpoints can be obtained.

1.2 开展管理人员现场安全督导工作(LSVP)的要求

1.2.1 参与方

(1) 高层管理人员

高层管理人员需要开展管理人员现场安全督导工作(LSVP)。他们每季度应至少完成一次管理人员现场安全督导。然而，根据企业业务情况的要求，高层管理人员也可以开展更多的管理人员现场安全督导(比如，由于生产作业场所发生HSE事件或HSE审核发现)。

高层管理人员应准备一个年度管理人员现场安全督导工作计划表，列明什么时间开展管理人员现场安全督导，督导哪些生产作业场所。管理人员现场安全督导计划表(示例)参见表1-1。督导计划表应随时更新，展示全年工作进展，识别必要的计划变更。

(2) 所有部门

各部门应确定哪个安全专题跟自己的业务活动相关。如必要，部门经理也可开展额外的安全专题，比如：其他部门的业务活动直接影响本部门的业务活动。

每个日历年，中层管理人员应完成其所在部门管理人员现场安全督导(LSV)的所有安全专题。其他部门成员完成哪些安全专题以及多少次管理人员现场安全督导(LSV)则视情况而定。

如果在一年中某个部门有跟特定安全专题相关的大量活动，部门经理可以决定针对特定安全专题开展多次管理人员现场安全督导(LSVs)。从有益于管理人员现场安全督导工作(LSVP)的角度来说，同一安全专题的不同次管理人员现场安全督导应由同一部门的不同人去完成，以获得不同的意见。

Table 1-1 Leadership Site Visit Matrix (Sample)

Position	Name	Jan	Feb	Mar	Apr	May	Jun	Jul	Aug	Sep	Oct	Nov	Dec
President													
Vice President													
……													
Manager													
Deputy Manager													
……													
Section Head													
……													

Instructions: 1. Include all LSV program participants in the matrix (position and name).

2. Complete the matrix by entering the topic number in the month in which it will be completed.

3. Topics should be varied for each month to prevent all participants completing the same Leadership Site Visit Prompt Card in the same month.

表 1-1　管理人员现场安全督导计划表(示例)

岗位	姓名	一月	二月	三月	四月	五月	六月	七月	八月	九月	十月	十一月	十二月
总经理													
副总经理													
……													
经理													
副经理													
……													
科长													
……													

注：1. 计划表应包含所有管理人员(岗位和姓名)。

2. 将每月计划督导的主题填入计划表中。

3. 不同管理人员在相同月份尽量选择不同的督导主题以避免在同一月份所有管理人员只督导了一个主题。

In order to assist in the planning of the site visits, each department shall produce a Visit Matrix of when the visits will be completed. The Visit Matrix shall be submitted to the HSE Department for the coming calendar year and shall be updated throughout the year to show progress and identify any necessary changes to the plan.

In order to prevent all topics being completed in a single "batch", and to ensure ongoing commitment to the program, a maximum of four topics shall be allowed each month. It is the responsibility of each Department Manager to ensure that all planned LSVs are completed as per the plan.

Although there is no requirement to conduct more LSVs than are identified in the annual visit matrix provided to the HSE Department, each Department is encouraged to perform additional LSVs when relevant situations arise. For example, the Construction Department may consider performing an LSV during a large lifting operation.

If visits are performed for topics not identified in the plan, the visit matrix shall be updated and re-issued to the HSE Department.

(3) Contractors

Contracting companies shall participate in the LSV program as appropriate to their activities.

1.2.2 Conducting the LSV

The LSV should be kept informal and relaxed as far as possible. This helps to prevent people becoming defensive and challenging and will promote the free exchange of information. As such, the LSV can be conducted as part of a normal visit to the site and does not need to be announced before hand.

为便于管理人员现场安全督导的策划，每个部门应制作一个管理人员现场安全督导计划表，表明什么时间完成什么专题的管理人员现场安全督导。每个部门在下一年的安全督导计划表应提前提交给 HSE 部门，并应及时更新，以展示管理人员现场安全督导工作的进展情况，识别必要的计划变更。

为避免在一次督导完成所有安全专题，确保管理人员现场安全督导工作的持续性，每月的管理人员现场安全督导工作不应超过 4 个安全专题。部门经理应切实履行其职责，确保按既定计划完成管理人员现场安全督导(LSVs)。

以提交给 HSE 部门的安全督导计划表为基准，虽然没有要求开展更多的管理人员现场安全督导(LSVs)，但相关情形出现时，鼓励每个部门开展更多的管理人员现场安全督导(LSVs)，比如：工程部门在大型吊装作业期间，可以安排一次针对性的管理人员现场安全督导(LSV)。

如果执行的管理人员现场安全督导(LSV)没有标注在督导计划中，安全督导计划表应适时更新并重新提交给 HSE 部门。

(3) 承包商

承包商也应参与到跟其作业活动相关的管理人员现场安全督导工作(LSVP)中。

1.2.2 开展安全督导

管理人员现场安全督导(LSV)应以尽可能轻松和非正式的方式进行。这有助于防止人们产生戒心和挑衅心理，并将促进信息的自由交流。这样，管理人员现场安全督导(LSV)可以变成普通现场访问的一部分，不需要提前通告。

(1) LSV Work Flow

In general, LSVs will be conducted in accordance with the following task list. However, the LSVP is intended to provide all Managers with flexibility to conduct the LSVs in such a way as to gain the most benefit (provided that the general theme of each visit is met).

- Prepare for visit:
  - *Visit when relevant task is ongoing.*
  - *Visit when relevant people are available.*
  - *Read relevant section of this set of books.*
- At site:
  - *Keep visits informal and relaxed.*
  - *Ask the "prompt" questions and follow-up with more detailed questions as necessary.*
  - *Ask additional questions where necessary.*
  - *Take notes of answers as necessary.*
  - *Conduct a Safety Conversation.*
  - *Identify corrective actions and close out if possible.*
  - *Record corrective actions that cannot be closed out immediately and will be registered in the LSVP Tracking System.*
- Post-visit:
  - *Complete the LSVP Tracking System with details of the visit, including any outstanding observations and the necessary corrective actions.*
  - *Follow-up on any actions to be performed by participants and close out in the LSVP Tracking System.*

(1) 管理人员现场安全督导(LSV)工作流程

通常，管理人员现场安全督导(LSVs)遵照以下任务清单予以开展。然而，管理人员现场安全督导工作(LSVP)允许所有管理人员保持灵活性，以获得最大的管理人员现场安全督导(LSVs)成效(只要满足每次安全督导大致的主题即可)。

- 督导准备
  - 根据正在进行的生产作业任务，计划督导时间。
  - 根据相关人员是否在场，计划督导时间。
  - 阅读本丛书中的相关部分。
- 督导现场
  - 以尽可能轻松和非正式的方式进行督导。
  - 如有必要，以管理人员现场安全督导提示卡片上的简要问题为开篇，探讨更详细的问题。
  - 如有必要，探讨更多的问题。
  - 如有必要，记录现场工作人员对问题的回应。
  - 跟现场工作人员进行安全交谈。
  - 识别纠正措施，实施并尽快关闭。
  - 将暂时无法关闭并将在管理人员现场安全督导工作(LSVP)追踪系统中进行登记的纠正措施记录下来。
- 督导后
  - 完成管理人员现场安全督导工作(LSVP)追踪系统所需信息，包括任何未完成的问题项以及必要的纠正措施。
  - 跟踪并实施本人需要采取的行动，最终在管理人员现场安全督导工作(LSVP)追踪系统中予以关闭。

(2) Preparing for the Visit

There should be little need for preparation as the visits are intended to be informal and straightforward. The visit can be conducted as part of the participant's normal work routine and specific formal site visits are not necessarily required.

While little preparation is needed, it would be useful for the person conducting the LSV to:

- Review the results of previous LSVs on the same topic across all sites in order to help identify company wide as well as site specific issues.
- Review all outstanding actions for the LSV topic in order to identify close-out opportunities.

(3) The Leadership Site Visit Prompt Card

This section is intended to provide instruction on how to complete the LSV Prompt Cards in order to obtain a minimum level of consistency by all participants. Specific topics at specific locations may need to be completed differently, however the overall principles of the LSV Prompt Cards should be maintained. It is anticipated that the participants will use their own knowledge and judgement in order to obtain the most appropriate information while on site.

Each safety topic should be provided with an LSV Prompt Card. The LSV Prompt Card is divided into four sections:

- **Questions:** The first page is the largest section and contains the questions specific to that topic.
- **Safety Conversation:** This section is for recording the findings of the Safety Conversation.
- **Notes:** A free space to add any relevant notes or to continue the answers to the questions.

(2) 督导准备

管理人员现场安全督导工作是非正式的和简单易懂的，因此并不需要过多准备。管理人员现场安全督导可以作为督导者正常工作的一部分，并不需要“过于正式”。

虽然管理人员现场安全督导(LSV)不需要过多事先准备，督导者考虑以下情形是有用的：

- 针对同一专题，回顾先前在不同生产作业场所开展管理人员现场安全督导(LSVs)的结果，以帮助识别特定场所的问题项以及本单位范围内的问题项。
- 回顾所有未关闭的安全督导(LSV)问题项，识别关闭机会。

(3) 管理人员现场安全督导提示卡

本节旨在提供有关如何完成管理人员现场安全督导提示卡的说明，以便所有督导者获得最低程度的一致性。特定地点的具体安全专题可能需要以不同方式完成，但应坚持管理人员现场安全督导提示卡的总体原则。督导者预期将使用他们自己的知识和判断，以便在现场获得最恰当的信息。

每个安全专题应有配套的管理人员现场安全督导提示卡，管理人员现场安全督导提示卡分为四个部分：

- **问题**：第一页是管理人员现场安全督导提示卡最大的部分，包含该安全专题特有的问题。
- **安全交谈**：这部分用于记录安全交谈的发现。
- **备注**：这是可以添加任何相关备注或继续记录问题回应的自由空间。

- **Visit Details:** This area is used to provide details of who conducted the visit, who was involved in the visit and when the visit was completed.

The best format for printing the LSV Prompt Card is A4, landscape. This provides the largest area to write in. In order to reduce the amount of paper used, it is suggested to print the form "double sided".

The purpose of the LSV Prompt Card is to provide a list of questions which should be asked as part of the LSV. These questions are prompts to begin discussion and should not be considered as stand-alone questions. The participant should ask additional questions as necessary.

The questions require an evaluation of site compliance to HSE-MS requirements and HSE work conditions. The evaluation uses a numerical rating system that measures the implementation/maturity of the HSE topic at the site being visited. This numerical system follows the guidance in Table 1-2.

Table 1-2 LSV Prompt Card Question Rating

Score	Documents	Implementation	Status
0	No Procedure No Record Keeping	No Understanding No Implementation	No HSE
1	Incomplete Procedure Incomplete Record Keeping	Incomplete Understanding Incomplete Implementation	HSE on Paper
2	Complete Procedure Incomplete Record Keeping	Complete Understanding Incomplete Implementation	HSE in Minds
3	Complete Procedure Complete Record Keeping	Complete Understanding Complete Implementation	HSE in Hearts and Minds

• **其他督导信息：**此区域用于提供督导者、参与督导工作的相关人员以及督导完成时间的详细信息。

打印管理人员现场安全督导提示卡的最佳方式是A4纸和横向版式，这提供了最大的书写区域。为了减少使用纸张的数量，建议“双面”打印。

管理人员现场安全督导提示卡的目的是为管理人员现场安全督导(LSV)提供一个问题列表，应作为现场安全督导的一部分探讨这些问题。它们是开始探讨的提示性问题，不应视为独立的问题。督导者应在必要时提出其他问题。

这些问题要求评估生产作业场所是否符合HSE管理体系(HSE-MS)要求和工作所需的HSE条件。评估使用量化评级系统，该系统测量所督导的生产作业场所特定安全专题的实施情况和成熟度。该量化评级系统遵循表1-2中的指引。

表1-2 督导提示卡问题打分指引

打分	文件	实施	状态
0	无程序 无记录	不理解 未实施	无安全
1	程序不完整 记录不完整	理解不完整 实施不完整	安全浮于表面
2	程序完整 记录不完整	理解完整 实施不完整	安全在心中
3	程序完整 记录完整	理解完整 实施完整	安全在心中并付诸行动

(4) Asking the Prompt Questions

The prompt questions should be asked as introductory questions and additional questions asked where necessary in order to understand a situation as completely as possible. Supporting information is provided in this set of books to assist participants in completing the questions.

Answers to the questions should be completed in the relevant section on the first page and in the notes section if necessary. Where significant information is required to answer a question, separate sheets may be used. However, the use of additional pages is envisaged to be the exception.

The prompt card questions are designed to uncover deficiencies in the systems being reviewed. All deficiencies (negative observations) should be recorded on the LSV Prompt Card and corrective actions shall be developed to resolve deficiencies and implemented as soon as reasonably practicable. It is expected that the majority of corrective actions are likely to be closed out during the visit.

The questions in the LSV Prompt Card should be assessed against the gudiance in Table 1-2. The score before and after corrective actions can be recorded on the LSV Prompt Card. It should be noted that any question where corrective action needs to be taken cannot score a rating of 3. This is due to lack of evidence to demonstrate that the system has been consistently implemented.

An informal Safety Conversation should be conducted with any non-senior person at the site in order to assess their understanding of the topic. This does not have to be someone in the same department, or the same company, but someone who does have interaction with the topic.

(4) 探讨管理人员现场安全督导提示卡上的问题

应将提示卡上的问题作为开篇，并在必要时提出其他问题，以便尽可能完整地了解情况。本丛书提供了支持性信息，以帮助督导者对问题项有深入理解。

如有必要，应在管理人员现场安全督导提示卡第一页的相关部分和备注(注释)部分填写对问题的回应。如果需要重要信息来回应问题，可以单独填写。但是，使用附加页被视作例外情况。

管理人员现场安全督导提示卡上的问题旨在揭示被督导生产作业场所在 HSE 风险防控方面的缺陷。所有缺陷(负面观察)都应记录在管理人员现场安全督导提示卡上，并应制定纠正措施以解决缺陷，在合理可行的范围内尽快实施。预计在督导期间可以关闭大部分纠正措施。

应根据表 1-2 中的指引评估管理人员现场安全督导提示卡中的问题。纠正措施前后的分数可记录在管理人员现场安全督导提示卡上。应该指出的是，任何需要采取纠正措施的问题都不能评为 3 分，这是因为没有证据证明所督导场所的 HSE 管理体系在一段时间内的实施是一致的。

跟生产作业场所的一般人员进行非正式的安全交谈，评估他们对某个安全专题的理解。这不意味着交谈对象必须是同一个部门或者同一个单位，只要跟某个安全专题有相互影响的人员即可。

The following is a quick guide to conducting a Safety Conversation. Detailed information and guidance on conducting Safety Conversations is provided in Section 1. 2. 3.

- Listen attentively.
- Emphasize positive actions you have observed.
- Do not apportion blame to individuals.
- Use language and questions that allow you to extract pertinent information from the person you are talking to.
- Get them to tell you, in their own words, what they ought to be doing in order to be safe.
- Ask questions with a sincere and caring demeanour.
- Act as if you do not know the answer, even though you think you do.
- Seek commitment to embrace and implement safety improvements.
- Agree on an action plan to achieve the safety improvements.

(5) Visit Details

The visit details must be record on the LSV Prompt Card. This allows tracking of which visits have been completed and when. It includes other people involved in the visit but do not name the person who had the Safety Conversation (unless they specifically request to be named).

(6) Post-Visit Tasks

All deficiencies that cannot be closed out before the end of the visit should be recorded in the LSVP Tracking System. This requires a recommended action to be developed and recorded and the implementation of that action to be assigned to a person. Where the deficiency is within the system rather than within the participant's department, these shall be elevated to Department Manager level and reported to the HSE Manager as appropriate.

以下是进行安全交谈的快速指引。关于安全交谈的详细信息和指引，参见本书 1.2.3 部分。

- 认真倾听。
- 强调你观察到的积极行动。
- 不要将责任归咎于个人。
- 使用合适的语言和问题，使得你可以从与你交谈的人那里提取切题的信息。
- 让与你交谈的人用自己的话告诉你，为了安全他们应该做些什么。
- 以真诚和关怀的举止提问。
- 即使你认为自己知道答案，也“假装”你不知道答案。
- 寻求接受和实施安全提升的承诺。
- 就行动计划达成一致以实现安全提升。

(5) 督导细节信息

必须完成管理人员现场安全督导提示卡要求填写的督导细节。这样可以跟踪已完成的督导次数和时间。这包括参与此次管理人员现场安全督导(LSV)的其他人员，但不要指出进行安全交谈的对象(除非他们特别要求写明)。

(6) 督导后的工作

在督导结束前无法关闭的所有 HSE 管理体系缺陷应记录在管理人员现场安全督导工作(LSVP)追踪系统中。这需要制定并记录建议的措施，并将该措施的实施分配给一个责任人。如果识别出的缺陷不在督导者所在部门内，则应将其上报至本部门经理，并适时向 HSE 经理报告。

Any actions identified and agreed during the Safety Conversation shall be followed up by the participant until closed out.

The LSV shall be recorded in the LSVP Tracking System even if no outstanding observations are recorded as the completion of LSVs is one of the program KPIs.

1.2.3 Safety Conversation

Quality Safety Conversations are one of the most effective tools we can use to create good communication flow between the workforce and all Managers and thereby create an environment where everyone values safety.

(1) What is a Safety Conversation?

A Safety Conversation is simply a conversation about safety. It can be a single topic or a general discussion. However, although it might sound simple, a Manager having a conversation with a worker, without sounding as if they are telling the worker off, or giving a lecture is a technique which needs to be learned.

The conversation should be anonymous to prevent the person worrying that there will be negative repercussions from them talking openly. This must be stressed at the beginning of the conversations, and can be repeated at any time during the conversation if the person appears to be uncomfortable and worried.

The Safety Conversation is designed to assess how that individual sees HSE matters in their work area, the site and across the company as a whole.

在安全交谈期间确定和同意的任何措施，应由督导者跟进直至关闭。

即使没有未关闭的待整改项，管理人员现场安全督导(LSV)也应记录在管理人员现场安全督导工作(LSVP)追踪系统中，因为每次管理人员现场安全督导(LSV)的完成情况是整个管理人员现场安全督导工作(LSVP)的关键绩效指标(KPIs)之一。

1.2.3 安全交谈

高质量的安全交谈是用来在现场工作人员和所有管理人员之间建立良好信息流动的最有效工具之一，并借此创建一个人人重视安全的环境。

(1) 什么是安全交谈?

简单来说，安全交谈是关于安全的对话。它可以是单个安全主题或一般性讨论。然而，虽然听起来可能很简单，但是当管理人员跟现场工作人员交谈时，管理人员听起来没有像在斥责人或者在做演讲，是一项需要学习的技能。

交谈应该是匿名的，以防止人们担心他们的坦诚交谈会产生负面影响。这一点必须在交谈开始时予以强调，如果交谈对象看起来不舒服或者有顾虑，可以在交谈期间的任何时间重复这一点。

安全交谈旨在评估个人如何看待 HSE 事项在其工作区域、整个生产作业场所以及整个工作单位的重要性。

(2) Why Have Safety Conversations?

Safety Conversations have many benefits with respect to creating a positive safety culture. Some of these include:

- **Preventing injuries and property loss.** The more safety is spoken about on the "shop floor", the less likely incidents occur or happen.
- **Reinforcing positive safety behaviour.** Safe behaviour is reinforced and unsafe behaviour is stopped.
- **Raising safety awareness in the workforce.** The more Safety Conversations that are held throughout the day the greater the safety awareness will be in the workforce.
- **Establishing standards.** Consistent communication of safety expectations assists in establishing standards to create a "this is the way things are done" environment.
- **Testing understanding of standards.** Safety Conversations provide a measure of how well standards are understood by the workforce.
- **Testing compliance with standards.** Being out on the "shop floor" provides an opportunity to gauge which standards are being followed and which are not.
- **Identifying weaknesses in safety systems.** Two way communication can lead the workforce to identify weaknesses and potential improvements in safety systems that can be fixed and actioned.
- **Identifying and correcting unsafe situations.**
- **Motivating people.** The workforce will generally feel more positive and empowered with respect to safety when they can see that Senior Managers are setting good examples. As a result, workers are more likely to continue positive safety behaviour.

(2) 为什么要进行安全交谈?

安全交谈在创建积极的安全文化方面有很多好处。其中一些好处包括：

- **预防伤害和财产损失**。“车间”或“班组”的安全性越高，安全事件发生的可能性就越小。
- **强化积极的安全行为**。安全行为得到加强，不安全的行为也会停止。
- **提高员工的安全意识**。每天开展的安全交谈越多，员工的安全意识就越强。
- **确立标准**。对安全期望的一致沟通有助于确立标准，借此创建一个“按标准做事”的环境。
- **测试对标准的理解**。安全交谈提供并衡量工作人员对标准的理解程度。
- **测试是否符合标准**。走进“车间”或“班组”提供了一个机会来衡量哪些标准被遵循，哪些标准没有。
- **识别 HSE 管理体系的弱项**。双向沟通可以引导工作人员识别可以修复和采取行动的 HSE 管理体系的弱项以及潜在改进机会。
- **识别和纠正不安全的状况**。
- **激励工作人员**。当工作人员看到高层管理人员树立良好榜样时，谈到安全通常会觉得更积极和更有力量。结果是工作人员更有可能继续积极的安全行为。

(3) Where to Have a Safety Conversation?

The Safety Conversation can be held in an office, at the work site or any other location, however the location should be one that allows the worker to feel relaxed. This is likely to be in locations such as the control room or the break room. Holding the conversation in a Manager's office should be avoided as this can be seen as a formal interview rather than an informal conversation.

(4) Who do I Have a Safety Conversation With?

Participants can have Safety Conversations with anybody. It would be normal for them to concentrate on people within their own department but there is no restriction on talking to people in other departments. If executed correctly and frequently, Safety Conversations can lead to improvements in the safety culture of the organisation.

Participants can also have conversations with contract staff. Helping to promote a positive attitude among contractors is very important.

In general, it is expected that Safety Conversations shall be held with supervisors and workers at the "shop floor" level. In some circumstances, particularly if the conversation is about HSE-MS policies and procedures, the conversation can be had with Middle Managers.

Where groups of workers do the same job (for example cleaning staff or general labourers) the Safety Conversation can be a group conversation. However, care must be taken to include all participants and not to let one person dominate the conversation.

(3) 在哪里进行安全交谈？

安全交谈可以在办公室、生产作业场所或任何其他地点进行，但是，该地点应该是允许工作人员感觉放松的地方，可能是在控制室或休息室等场所。应尽可能避免在管理人员办公室进行安全交谈，因为这可以被视为正式访谈而不是非正式交谈。

(4) 跟谁进行安全交谈？

督导者可以与任何人进行安全交谈。他们专注于自己部门内的人员是正常的，但并没有限制跟其他部门的人员交谈安全问题。正确且经常性地开展安全交谈，可以改善组织的安全文化。

督导者还可以与承包商工作人员进行交谈，帮助促进承包商对 HSE 的积极态度非常重要。

一般而言，安全交谈的对象是生产作业场所的主管人员及一般工作人员。在某些情况下，特别是如果交谈 HSE 管理体系(HSE-MS)政策和程序时，交谈的对象可以是中层管理人员。

如果一组工作人员从事相同的工作(例如保洁人员或普通力工)，则安全交谈可以是群体形式的交谈。但是，必须注意包含所有参与者，而不是让一个人主导整个交谈。

(5) How to Conduct a Safety Conversation?

"*The art of good conversation lies in the ability to listen*"——Malcolm Forbes (publisher of Forbes Magazine).

It is important to make sure that the other person feels relaxed and happy to talk. Listening to them more than you talk will help them feel you care about what they have to say. If they feel threatened, or if they feel they will get into trouble, they will not want to talk and will not be open and honest.

The following are tips for conducting good Safety Conversations:

- Maintain a positive attitude towards safety and towards Safety Conversations. People quickly sense insincere and negative attitudes.
- Prepare your opening line and start off with a positive statement. If you notice something good, discuss it.
- Frame the conversation with care and concern and the discussion will more likely be interactive. Ask questions in a non-threatening way.
- Do not influence the Safety Conversation with negative feelings towards something or somebody. Keep positive towards safety but neutral towards other issues.
- Do not let your prejudices about a speaker limit what you hear. Tell yourself you are not listening to someone, rather you are listening for something. You are not listening reactively to confirm a prejudice——you are listening proactively for clues or possibilities to solve a safety issue.

(5) 如何进行安全交谈?

良好交谈的艺术在于倾听能力——马尔科姆·福布斯(福布斯杂志的出版人)。

确保你的交谈对象感到放松并高兴谈话，这一点是重要的。多听少讲，将帮助他们感觉你在意他们要说的内容。如果他们感觉受到威胁，或者是如果他们感觉他们将陷入麻烦，他们会不愿意讲话，并且不会开放和诚实。

以下是进行良好安全交谈的一些小窍门：

- 对安全和安全交谈保持积极的态度。人们会快速感知到不真诚和负面态度。
- 准备你的开场白，从积极的内容开始。如果你注意到良好的事项，对其进行讨论。
- 谨慎对待交谈，讨论更有可能是互动的，以“非威胁”的方式提问。
- 不要以对某事或某人的负面情绪影响安全交谈。对安全问题保持积极态度，但对其他问题保持中立。
- 不要让你对发言者的偏见限制你听到的内容。告诉自己你不是在听某人，而是在听某事。你没有被动地倾听以确认偏见——你正在积极地寻找解决安全问题的线索或可能性。

• Be aware of cultural differences and the Safety Culture context of the Safety Conversation. Culture greatly influences the way people think and act. The aim is to help persuade them to change, not force them.

• Observe what people do and how they do it but do not make them think you are trying to catch them out.

• Watch body language and listen to how they speak. If they are tense and nervous you need to make them relaxed and at ease.

• Learn how to tactically ignore. If something is minor ignore it and focus on the major issues.

• Look out for blaming, rushing, distractedness, fatigue, arrogance, overconfidence and know how to soften and reframe dangerous language.

• Be focused on talking and guiding, not "telling" people how it should be done.

• Ask open questions. Avoid questions with yes/no answers.

• Do not ask confirmatory or leading questions, you will make people answer what you (they think) want to hear.

• Do not ask questions which try to prove your own assumption. If they have a theory, they should be able to tell it.

• Ask questions which help them tell a sequence and story.

• Ask creative questions. Questions which may not be exactly "on subject" but allow you to open up the conversation to different ideas and topics.

• Use examples from your past but do not concentrate solely on your own experiences. The conversation is two way and the experiences of the person you are talking to may be very relevant.

• 意识到安全交谈的文化差异以及安全文化方面的语境。文化极大地影响着人们的思考和行为方式。目的是帮助说服他们改变，而不是强迫他们。

• 观察人们做了什么以及他们如何做，但不要让他们认为你正试图挑他们的毛病。

• 观看肢体语言并倾听他们的说话方式。如果他们紧绷和紧张，你需要让他们松弛和放松。

• 学习有策略地忽略。请忽略次要问题，专注于主要问题。

• 留神语言中包含的责备、冲动、分心、疲劳、傲慢、过度自信，并知道如何缓和和重构危险的语言。

• 专注于谈论和引导，而不是“告诉”人们应该如何做。

• 提出开放式的问题。避免提出答案为“是/否”的问题。

• 不要提出确认式或主导式的问题，这只会让人们回应他们认为你想听到的内容。

• 不要提出试图证明自己假设的问题。如果他们有推测，他们应该能够说出来。

• 提出有助于他们讲述片段和故事的问题。

• 提出有创意的问题。问题可能不完全符合交谈“主题”，但允许你打开不同想法和主题的对话。

• 使用你过去的例子，但不要只聚焦你自己的经历。交谈是双向的，你正在交谈的人的经历可能跟交谈内容非常相关。

- Understand how to end the conversation. End on a positive note. Make the other person feel that their opinion is valued.
- Thank them for their time. If you have said you will do something then reiterate it so they know you have remembered.

(6) What to Avoid in a Safety Conversation?

Safety Conversations can be a very potent tool in creating a positive safety culture, but if done poorly can actually do more harm than good. Some common mistakes that people make when undertaking a Safety Conversation include:

- Conducting a planned Safety Conversation when you do not have a positive attitude. People can sense authoritarian and negative attitudes.
- Having a set agenda of what you want or what you are looking for. If you are focused on one particular area then it is likely you may miss important information whilst you are having the conversation.
- Monitoring workers as part of a fault finding mission. This shows a lack of trust that is guaranteed to deteriorate the safety culture.
- Bringing out checklists and ticking it off whilst having the conversation.
- Only focusing on things the workforce were doing incorrectly. It is more important to focus on the safe behaviours and ensure they are appreciated for their efforts.
- Do not focus on small details. Focus on the big picture.
- Instructing people on how to improve. The purpose is to promote conversation and allow the workforce to come up with the ideas on how to improve safety.

• 知道如何结束交谈。以积极的方式结束，让对方感受到他们的意见是被重视的。

• 感谢他们花费时间。如果你说过你会做某事，那就重申一下，这样他们就知道你已经记住了。

(6) 安全交谈应避开什么？

在创建积极安全文化方面，安全交谈是强有力的工具。但是，如果安全交谈做得不好，则可能得不偿失。人们在安全交谈时常犯的错误包括：

• 当你没有积极的态度时开展有计划的安全交谈。交谈对象可以感知到威权主义和负面的态度。

• 对于你要什么或找什么，有一个预设的一览表。如果你聚焦在一个特别的领域，你在安全交谈过程中可能错失重要的信息。

• 以发现缺陷为目的观察工作人员。这显示缺乏信任，肯定会破坏安全文化。

• 在安全交谈时，取出检查表，并在检查表上做标记。

• 只关注工作人员做得不对的事项。实际上，聚焦安全的行为更为重要，确保工作人员在安全行为方面的努力得到肯定。

• 只关注小细节。实际上，应该关注主要部分。

• 指导人们如何改进工作。实际上，安全交谈的目的是促进交谈，允许工作人员就如何提升安全拿出自己的主意。

- Asking closed questions.
- Asking confirmatory or leading questions.
- Asking questions which try to prove your own hypothesis or assumptions.
- Dominating the conversation with personal experiences. Conversations are not about you. Dialogue needs to be two way.

1.2.4 References

(1) http://www.hse.gov.uk/involvement/inspections.htm.

(2) Institute of Public Affairs (IPA), Australia, "Starting a Safety Conversation: Why they're important and how to have them".

1.3 Management Walkthroughs, Management Inspections and Management Audits

There are a number of different methods which can be used for managers to review the implementation of the HSE-MS at operational facilities. These fall into three broad categories:

1.3.1 Management Walkthroughs

There are a number of situations when Management Walkthroughs are carried out:

- By Middle Managers as part of normal routine business.
- By Senior Managers when they want to have a better understanding of site activities (often during shutdowns, after incidents, etc.) and to see how HSE matters are being managed on site.

- 询问封闭式问题。
- 询问确认式问题或主导式问题。
- 询问问题，试图证明你自己的假设或假定。
- 用个人经验控制整个交谈。实际上，交谈不只是你自己。对话需要双方向(互动)。

1.2.4 参考资料

(1) http://www.hse.gov.uk/involvement/inspections.htm.

(2) Institute of Public Affairs (IPA), Australia, "Starting a Safety Conversation: Why they're important and how to have them".

1.3 管理巡查、管理层检查和管理审核

有许多不同的方法可供管理人员审查 HSE 管理体系(HSE-MS)在生产设施中的实施情况。这些方法分为三大类：

1.3.1 管理巡查

管理巡查发生在许多情况下：

- 由中层管理人员作为其日常业务的一部分；
- 高层管理人员希望更好地了解现场活动(通常在停车期间、事故发生后等)，并了解现场如何管理 HSE 事项；

- During formal site visits by Very Important Person (VIP) visitors such as Government agencies, managers from other companies, etc., the VIP visitors will generally be escorted by a Senior Manager.

A walkthrough by Senior Managers and VIP visitors will likely be planned in advance. Walkthroughs by Middle Managers are unlikely to be planned.

A walkthrough is a quick, high level visit to the facility. The Manager (Senior or Middle) and the VIP visitor if in attendance, will walk through the site, visit specific locations (such as control rooms, fire stations, etc.) and generally walk around looking at areas and activities. All managers would be expected to directly question involved personnel where they see conditions which they believe may contravene HSE rules.

It is normal for issues such as housekeeping, wearing of PPE, waste segregation, etc. to be highlighted by a walkthrough. Issues of this nature can generally be corrected during the visit.

Depending upon the manager, whether Senior or Middle, they may decide to look at something in greater detail, particularly if they find multiple infringements on the same visit. The manager may also decide to look at certain equipment or certain activities in detail on the visit (the subject is likely to be connected with work they have done in the past or where there have been previous issues on this site or other sites the manager has worked on or visited).

• 在贵宾(VIP)访客(如政府机构、其他企业的管理人员等)正式实地考察期间，贵宾访客通常由本单位的高层管理人员陪同。

高层管理人员和贵宾(VIP)访客的管理巡查多数会提前计划。中层管理人员的管理巡查不一定提前计划。

巡查是对运营设施快速、程度较浅的访问。管理人员(高层管理人员或中层管理人员)和贵宾访客如果进行管理巡查，将在生产作业场所中走走停停，访问特定地点(如控制室，消防站等)，在走动中查看某些区域和某些活动。当看到生产作业场所中某些人员可能违反 HSE 规定的情况时，所有管理人员预期将直接询问这些人。

管理巡查通常关注生产作业场所是否整洁有序、现场人员是否正确穿戴了个体防护用品(PPE)、废物是否得到正确分离等事项，在访问期间就可以纠正这类问题项。

高层管理人员或中层管理人员可能会决定更详细地查看某些内容，特别是当他们在同一次访问中发现多次违章行为的时候。管理人员还可能决定在访问时详细查看某些设备或某些活动(决定可能跟他们过去所做的工作、本场所先前存在的问题项、他们本人工作过或访问过的其他场所存在的问题项等有关联)。

Detailed interaction between Senior Managers and personnel at site is likely to be limited to Site Level Managers. Interaction between Senior Managers and site personnel at a "shop floor" level will likely be limited to generic activity questions or generic safety questions.

No formal documentations need to be developed for a walkthrough, however, if issues are found during the walkthrough, these should be relayed to the respective Department Heads.

Depending on the size of the facility, a walkthrough of the operational facilities could take about one hour per site.

1.3.2 Management Inspections

Management Inspections are more in-depth than a walkthrough and would generally be completed by the Senior Management team. These would be completed on site, and often be linked to a specific topic or activity. They are often performed after an incident at the facility or where an incident has happened at a similar facility.

The Senior Managers would be interested to find specific issues surrounding the topic and attempt to determine why the incident occurred. Interaction with site personnel would be formal but records are not likely to be kept.

The Senior Manager should be specific about the topic to be reviewed, but ideally, the visit to site would start with a quick site walkthrough.

高层管理人员与现场人员之间的详细互动可能仅限于现场管理人员。高层管理人员跟现场一般人员的互动可能仅限于一般性活动的问题和一般性安全问题。

管理巡查不需要准备正式的报告。但是，如果在管理巡查期间发现了问题，这些问题应转发给问题所属部门的负责人。

根据设施的规模，针对每个站点生产设施的管理巡查可能需要 1h 左右。

1.3.2 管理层检查

管理层检查比管理巡查更深入，通常由高层管理人员组成的团队完成。管理层检查通常与特定主题或活动相关联，并将在现场完成。它们通常在生产作业场所发生 HSE 事件后，或在类似生产作业场所发生 HSE 事件后进行。

高层管理人员对找出特定主题的具体问题感兴趣，并试图确定 HSE 事件发生的原因。与现场人员的互动将是正式的，但不太可能保留记录。

高层管理人员应该具体讨论要审查的特定主题。但理想状况下，访问生产作业场所将从一个快速的管理巡查开始。

Information will be requested by the Senior Manager around the topic in question. The Senior Manager should review the detail and determine what additional information is required. If a problem is identified the Senior Manager may then request a formal audit of the topic. The formal audit could be conducted by a team with relevant auditing experience and subject knowledge or by a specialist third party, depending upon the requirements of the audit.

Although Senior Managers are not limited to which people they talk to at site, and interaction with personnel at all levels is encouraged. It is recognised that time constraints may limit interaction between Senior Managers and site personnel, particularly at a "shop floor" level, to determine how a topic is implemented or to gain information around an incident.

There is unlikely formal written reports to be produced following a Management Inspection. If a written report needs to be written, it would likely be short and would list deficiencies found. It is now common for the results of a Management Inspection to be reported as an e-mail. Deficiencies found would generally be included on the HSE Action Tracking system but are the responsibility of the respective department to close out the action.

If the Senior Manager has a specific idea regarding a corrective action, this may be included with the e-mail. However, it is more likely that the corrective actions will be developed by the relevant department and presented back to the Senior Manager.

Depending on the size of the facilities, a Management Inspection would be expected to take a number of hours rather than a number of days.

检查时，高层管理人员应要求就相关主题提供信息，审查细节并确定需要哪些附加信息。如果发现问题，高层管理人员可要求对该主题进行正式审核。正式审核可由具备相关审核经验和专业知识的团队进行，或由第三方专家根据审核要求进行。

虽然高层管理人员不会受限于在现场跟谁交谈，并且鼓励其与各级人员进行互动。为了确定某个主题的实施情况或获取事件有关信息，时间的限制可能制约高层管理人员和现场人员之间的互动，特别是在普通工人层面。

管理层检查不可能提供正式的报告。如果需要写一份报告，它可能很简短且只列发现的缺陷。现在较常见的做法是将管理层检查结果以电子邮件的形式提交。管理层检查发现的缺陷通常会包含在 HSE 措施追踪系统中，但相关部门有责任关闭缺陷纠正措施。

如果高层管理人员对缺陷纠正措施有明确的想法，这可以包含在电子邮件中，但更合适的做法是由责任部门制定纠正措施并呈报给提出问题的高层管理人员。

根据设施规模大小，管理层检查预计需要数小时而不是数天。

1.3.3 Management Audits

Management Audits are not necessarily audits completed by Managers. They would generally be completed by a team of people (led by a Lead Auditor) and would be expected to take a few days to complete. Interaction with personnel at site would be formal and records would be kept. The audit would generally look at the system as a whole, not at individual topics. However, topics may be selected for greater scrutiny if there have been issues in the past or if weaknesses are found in the documentation.

For companies, where their operations are located across multiple sites, it would be expected that the overall system would be audited at the administration centre (Main Base Camp). Verification of the implementation of the system at the sites would be completed and would likely concentrate on areas of concern within the overall audit.

Management audits would not necessarily go into great depth about the quality of the systems, although improvement actions would be expected. Audits of this sort tend to concentrate on the implementation of the system and so would be looking at records (such as PTW, training, etc.) and would try to identify where HSE improvements (or otherwise) have been made since the last audit.

The output from this type of audit would be a report detailing the findings and suggesting some improvement actions for the overall system, specific system documents or the implementation of the system, as appropriate.

These audits should not be confused with very in-depth System Audits (such as ISO 9001, ISO 14001 or ISO 45001). These audits would be expected to take a number of weeks and would be very in-depth with significant deliverables.

1.3.3 管理审核

管理审核不一定是由管理人员完成的审核，通常由主任审核员带领一组人来进行，预计需要几天时间才能完成。与现场人员的互动将是正式的，并保留记录。审核通常会考虑体系总体，而不是单个主题。但是，如果过去存在问题或者文档中存在缺陷，则可以选择具体主题进行更详细的审查。

对于生产作业位于多个站点的单位，预计体系总体将在其管理总部（主营地）进行审核，并将对体系在生产作业场所的实施情况进行验证，验证可能集中在总体审核需关注的领域。

虽然预期会采取改进措施，但管理审核不一定会对体系质量有深入审查。这种审核倾向于集中在体系实施情况，因此将查看记录[例如：作业许可（PTW）、培训等]，并试图确定自上次审核以来HSE改进（或其反面）的情况。

此类审核的结果将是一份详细说明审核发现的报告，并为体系总体、具体体系文件或体系实施提出一些适当的改进措施建议。

不应将这些管理审核与非常深入的体系审核（例如ISO 9001，ISO 14001或ISO 45001）相混淆。这些审核预计需要数周时间❶，审核非常深入且有丰富的交付成果。

---

❶具体时间长短取决于受审核单位的规模大小。

2

HIERARCHY OF CONTROLS

（危害）控制层序

2.1 Introduction

Controlling exposures to occupational hazards is the fundamental method of protecting workers. Traditionally, a hierarchy of (hazard) controls has been used as a means of determining how to implement feasible and effective control solutions. It is a widely accepted system promoted by numerous international safety organizations. This concept is taught to managers in industry and promoted as standard practice in the workplace. Various illustrations are used to depict this system, most commonly a triangle (Figure 2-1, NIOSH). Figure 2-1 shows the order to follow when planning to reduce occupational health and safety risk from work activities.

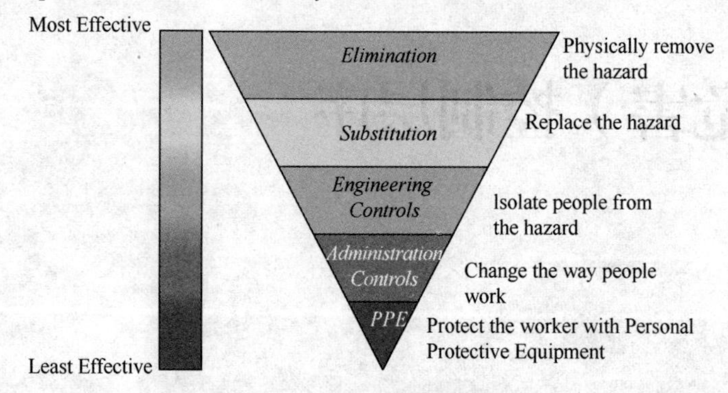

Figure 2-1 Hierarchy of Controls

The hierarchy of control shows the preferred order of dealing with hazards, with elimination/substitution being the most preferred and PPE the least. Starting from the top, move down the hierarchy so that if step 1, elimination/substitution, cannot be implemented, then try and implement step 2, isolation/separation, etc.

2.1 简介

控制职业危害暴露是保护工作人员健康安全的根本方法。一直以来，（危害）控制层序被用作确定如何实施可行和有效的危害应对措施的方法。（危害）控制层序是被众多国际安全机构所推崇并被广泛接受的体系。这种思路被讲授给工业中的管理人员并被推广为工作场所的标准做法。用于描述这个体系的插图很多，最常见的是三角形，如图 2-1 所示（NIOSH，美国国家职业安全卫生研究院）。图 2-1 列出了在策划降低来自工作活动的职业健康和安全风险时应遵循的先后顺序。

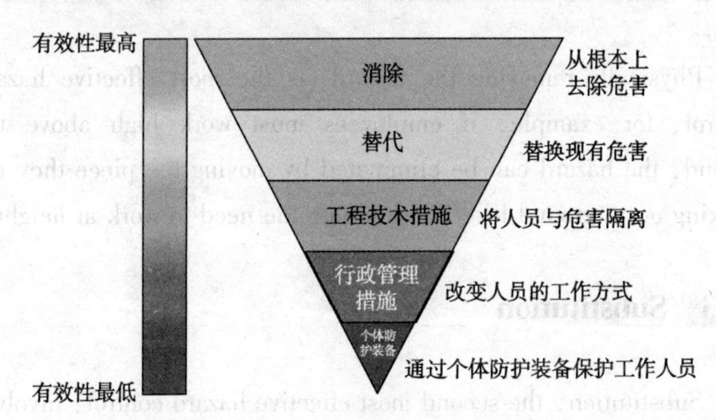

图 2-1 （危害）控制层序

（危害）控制层序展示了危害防控措施的分层和排序要求：最先考虑消除/替代措施，最后考虑个体防护装备。从层序图（图 2-1）的顶部开始向下，如果第一步（消除/替代）不能实施，那么就尝试和实施第二步（隔离/分离），以此类推。

The control measures should be considered in the order shown in Figure 2－1, and the temptation to jump to the easiest control measure to implement should be avoided.

The hierarchy of (hazard) controls is a core component of Prevention through Design, the concept of applying methods to minimize occupational hazards early in the design process. Prevention through Design emphasizes addressing hazards at the top of the hierarchy of controls (mainly through elimination and substitution) at the earliest stages of project development.

2.2 Elimination

Physically removing the hazard－is the most effective hazard control, for example, if employees must work high above the ground, the hazard can be eliminated by moving the piece they are working on to ground level to eliminate the need to work at heights.

2.3 Substitution

Substitution, the second most effective hazard control, involves replacing something that produces a hazard (similar to elimination) with something that does not produce a hazard, for e. g., replacing lead-based paint with titanium white. To be an effective control, the new product must not produce another hazard, For example, where airborne dust is a potential hazard, an effective control measure is the substitution of the smaller particle size product with one of a larger particle size, if available.

应按图 2-1 展示的顺序考虑(不同有效性的)防控措施，避免试图跳到最简单易行的防控措施。

(危害)控制层序是“设计保障安全”这一理念的核心组成部分，即在设计过程早期就采取措施尽量减少职业危害。“设计保障安全”强调在项目开发的最初阶段通过(危害)控制层序上层的方法控制危害(主要通过消除和替代)。

2.2 消除

完全去除危害是最有效的危害防控措施。例如：如果员工需要在高于地面的位置工作，则可以将工作对象移至地面来消除高处作业的危害。

2.3 替代

替代是有效性位列第二的危害防控手段。替代方法类似于消除，是将产生危害的物料替换为不会产生(同样)危害的物料，例如：用钛白涂料替代铅基涂料。为了确保防控措施的有效性，新物料也不能产生其他危害，比如，由于空气中的粉尘是危险有害的，如果可以购买更大粒径的产品，较小粒径的产品可被较大粒径的产品有效替代。

2.4 Engineering Controls

The third most effective means of controlling hazards is engineering controls. Engineering controls are designed to protect workers from hazardous conditions by placing a barrier between the worker and the hazard or by removing a hazardous substance through air ventilation. Engineering controls involve a physical change to the workplace itself, rather than relying on workers' behaviour or requiring workers to wear protective clothing.

2.5 Administrative Controls

Administrative controls are changes to the way people work. Examples of administrative controls include procedure changes, employee training, and installation of signs and warning labels. Administrative controls do not remove hazards, but limit or prevent people's exposure to the hazards, such as completing road construction at night when fewer people are driving.

2.4 工程技术措施

工程技术措施是有效性位列第三的危害防控手段。工程技术措施旨在通过在工作人员与危害之间设置屏障或通过通风去除有害物质来保护工作人员免遭危险状态的伤害。工程技术措施依靠工作场所本身的物理变化，而不是依赖工人的行为或要求工人穿防护服。

2.5 行政管理措施

行政管理措施主要依靠调整工作方式，例如修订程序、培训员工以及设置标识和警告标签。行政管理措施不能消除危害，而是限制或防止人员暴露于危害之中，例如选择在车流量较少的夜间完成道路施工作业。

2.6 Personal Protective Equipment

Personal protective equipment (PPE) includes gloves, Nomex Uniform, respirators, hard hats, safety glasses, high-visibility clothing, and safety footwear. PPE is the least effective means of controlling hazards because of the high potential for damage to render PPE ineffective. Additionally, some PPE, such as respirators, increase physiological effort to complete a task and, therefore, may require medical examinations to ensure workers can use the PPE without risking their health.

Elimination and substitution, while most effective at reducing hazards, also tend to be the most difficult to implement in an existing process. If the process is still at the design or development stage, elimination and substitution of hazards may be inexpensive and simple to implement. For an existing process, major changes in equipment and procedures may be required to eliminate or substitute for a hazard.

Engineering controls are favored over administrative and personal protective equipment (PPE) for controlling existing worker exposures in the workplace. Well-designed engineering controls can be highly effective in protecting workers and will typically be independent of worker interactions to provide this high level of protection. The initial cost of engineering controls can be higher than the cost of administrative controls or PPE, but over the longer term, operating costs are frequently lower, and in some instances, can provide a cost savings in other areas. For example, a crew might build a work platform rather than purchase, replace, and maintain fall arrest equipment.

2.6 个体防护装备

个体防护装备(PPE)包括手套、Nomex(一种防火防静电的材料)工服，呼吸器、安全帽、护目镜、高亮警示服和安全鞋等。由于损坏导致个体防护装备(PPE)功能失效的概率很高，所以个体防护装备(PPE)是有效性最低的危害防控手段。此外，某些个体防护装备，例如，呼吸器会增加使用者完成工作任务的生理负荷，因此对使用某些特殊个体防护装备(PPE)的人员可能要进行职业健康检查，以保证这些个体防护装备(PPE)不会对使用者带来健康风险。

尽管消除和替代在减少危害方面最为有效，但对在役工艺来说，往往被认为是最难实施的方法。如果在设计或开发阶段，消除和替代可能是经济且容易实施的。对在役工艺，往往需要对设备和程序进行重大变更才可能消除或替代现有危害。

为了有效防控人员暴露于工作场所的危害，工程技术措施往往优于行政管理措施和个体防护装备(PPE)。设计良好的工程技术措施可以非常有效地保护工作人员，并且通常不依赖于工作人员的交互作用就能提供高水平的危害防护。尽管工程技术措施的初期花费可能比行政管理措施或个体防护装备昂贵，但从长远来看，运行成本通常较低，并且在某些情况下，可能在其他方面节约成本。例如，如果建立了工作平台就不需要再购买、更换和维护防坠落装备。

Administrative controls and PPE are frequently used with existing processes where hazards are not particularly well controlled. Administrative controls and PPE programs may be relatively inexpensive to establish but, over the long term, can be very costly to sustain. These methods for protecting workers have also proven to be less effective than other measures, requiring significant effort by the affected workers.

2.7 References

(1) NIOSH, USA, Hierarchy of Controls.
https://www.cdc.gov/niosh/topics/hierarchy/default.html.
(2) Hierarchy of hazard controls.
https://wikivisually.com/wiki/Hierarchy_of_hazard_controls.

行政管理措施和个体防护装备(PPE)通常应用于危害没有被有效控制的现有工艺过程。行政管理措施和个体防护装备(PPE)的初始投入费用相对较少，但从长远来看，维护费用可能非常昂贵。这两种保护工作人员的措施证明不如其他措施有效，需要工作场所受影响人员相当多的努力才能奏效。

2.7 参考资料

(1) NIOSH, USA, Hierarchy of Controls.
https://www.cdc.gov/niosh/topics/hierarchy/default.html.
(2) Hierarchy of hazard controls.
https://wikivisually.com/wiki/Hierarchy_of_hazard_controls.

3

LINE OF FIRE

危险轨迹

3.1 Introduction

Technically, "Line of Fire" is a military term that describes the path of a discharged missile or firearm. Here "Line of Fire" refers to the path an object will travel creating risk of serious injury.

In many of the tasks that are performed on a daily basis, there is often the possibility of putting yourself in the line of fire. Statistics from the UK show that 78% of all work related fatal injuries were associated with some form of line of fire incident.

In order to avoid injury, it is important to understand what and where the line of fire is and how to keep out of it. Industry standards break line of fire into three mechanisms of injury (Table 3-1).

Table 3-1 Three mechanisms of injury

	**1. Stored Energy:** Contact with stored energy including pressure, gravity, objects under mechanical tension, etc. Stored energy is "pent up" energy that can be released unexpectedly.
	**2. Striking Hazards:** Struck by or striking against an object. Includes falling/dropped objects and moving equipment and vehicles.
	**3. Crushing Hazards:** Caught in, on or between an object.

3.1 简介

从技术角度来讲，“Line of Fire”是一个军事术语，用以描述导弹或子弹发射后的轨迹(称作“弹道”)。本书中的“Line of Fire”指的是物体将行进并(在行进过程中)产生严重伤害风险的路径(称作“危险轨迹”)。

在每天执行的许多任务中，经常有可能将自己置于“危险轨迹”当中。来自英国的统计数据显示，所有与致命伤害相关的工作中有78%与某种形式的“危险轨迹”事件有关。

为了避免受伤，重要的是要了解“危险轨迹”含义和位置以及如何置身“危险轨迹”之外。行业标准将“危险轨迹”分为三种伤害机制(表3-1)。

表3-1 三种伤害机制

	**1. 蓄能危险**：接触储存的能量，包括压力、重力，机械张力下的物体等。储存的能量是“被封存”的能量，可能意外释放
	**2. 撞击危险**：被物体打击或撞上物体。包括坠落物及移动设备和车辆
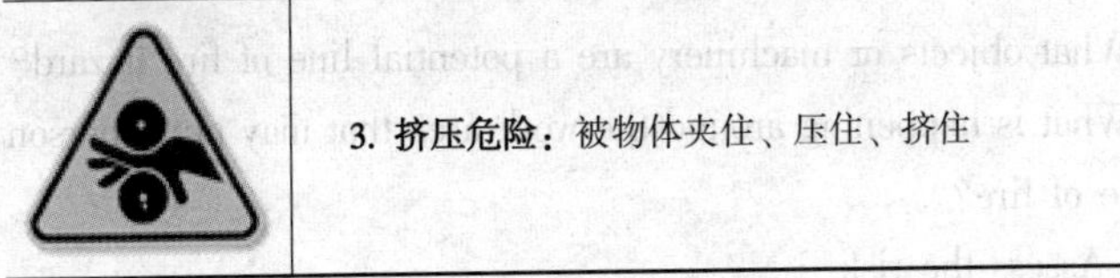	**3. 挤压危险**：被物体夹住、压住、挤住

Line of Fire is one of the new Life – Saving Rules being introduced by the International Association of Oil & Gas Producers (IOGP) in 2018.

3.2 Identification of the Line of Fire Hazards

"Awareness is our first line of defence" and as such, it is important that line of fire hazards are included in risk assessments conducted at a work site. Failure to identify these hazards and implement suitable controls can result in injury or worse. Ultimately, it would be preferable to eliminate the hazard, however, this is not always possible and as such, controls need to be identified and implemented to minimise the hazard.

It is important in risk assessment to consider the entire duration for the work being performed. It is easy to concentrate the risk assessment on the beginning of the job, making sure sufficient controls are in place to start the work. Hazards can change as the work progresses, or new hazards can be introduced, not just from the work being performed but from the surrounding area, weather, light levels, etc.

The following basic steps should be taken when considering line of fire hazards:

(1) Identify the line of fire hazards

- What objects or machinery are a potential line of fire hazard?
- What is happening around the work area that may put a person in the line of fire?

(2) Assess the risk

- If the line of fire hazard is released, are people likely to be in the line of fire?

危险轨迹是国际油气生产商协会(IOGP)于 2018 年推出的新的保命法则之一。

3.2 识别产生“危险轨迹”的危害

“意识是我们的第一道防线”，因此在工作现场进行的风险评估中将产生“危险轨迹”的危害包括在内是很重要的。未能识别产生“危险轨迹”的危害及实施适当的控制措施，可能导致伤害或更严重的情况。最终，消除危险是最可取的控制措施。然而，这并不总是可行的。因此，需要确定和实施控制措施，以尽可能降低危险。

在风险评估中，考虑正在进行的工作的整个持续时间是很重要的。很容易将风险评估集中在工作的开始阶段，确保有足够的控制措施来开始工作。但随着工作的进展，危害可能会发生变化，或者可能引入新的危险，不仅仅是来自正在进行的工作，还来自周边、天气、光照等。

在考虑产生“危险轨迹”的危害时，应采取以下基本步骤：

(1) 识别产生“危险轨迹”的危害

- 哪些物体或机器是潜在的产生“危险轨迹”的危害？
- 工作区周围正在发生什么事可能会使一个人陷入“危险轨迹”？

(2) 评估风险

- 如果危害沿“危险轨迹”释放，人员是否可能处于“危险轨迹”中？

- Be aware of where people are located throughout the entire task in relation to the direction of the hazardous energy (travel of a vehicle, whip action of a released pipe, etc.).

(3) Identify suitable controls

- Eliminate the hazard where possible (relocate vehicles, stop rotating machinery, etc.). If eliminating the hazard is not practicable, minimise the hazard using the hierarchy of control.
- Substitute the line of fire hazard with something safer.
- Engineering controls: Implement protection barriers such as guarding or de-energise line of fire hazards through isolation (lock out/tag out)
- Administration controls: Prohibit persons working in the line of fire through adequate supervision or developing exclusion zones. During work execution, position the body (specifically hands and arms) clear from the line of fire.
- PPE: Use adequate PPE such as gloves, hard hats and eye protection.

Although the main focus of identifying line of fire hazards is to protect people from injuries and fatalities, it should be recognised that line of fire hazards also apply to plant and equipment. This is particularly important if damage to a piece of equipment could cause a more hazardous scenario to occur. For example, process equipment struck by a vehicle can cause a loss of containment, or an electrical power transformer struck by a falling crane could cause the crane to become live and may cause power outages in the facility. It is important to include critical equipment in any line of fire risk assessment.

• 了解人员在整个任务中与危险能量释放方向(车辆行驶、管线末端剧烈摆动等)有关的位置。

(3) 确定合适的控制措施

• 尽可能消除危害(重新安置车辆、停止旋转机械等)。如果消除危害不可行,按照(危害)控制层序的原则尽可能降低危险。

• 用更安全的东西替代产生“危险轨迹”的危害。

• 工程控制措施:实施保护屏障(如防护罩)或通过隔离(挂锁/上签)切断产生“危险轨迹”的能量危害。

• 行政管理措施:通过恰当监管或设立禁止区[1],禁止人员在“危险轨迹”工作。在执行工作期间,确保身体(特别是手和手臂)不进入“危险轨迹”。

• 个体防护装备:使用恰当的个体防护装备,如手套、安全帽和护目镜。

虽然识别产生“危险轨迹”的危害的主要焦点是保护人员免受伤害和死亡,但应该认识到产生“危险轨迹”的危害也适用于设施和设备。如果对设备的损坏可能导致更危险的情况发生,这一点尤为重要。例如,车辆撞击工艺设备可能导致工艺泄漏,倾倒的起重机撞击电力变压器可能导致起重机带电并且可能导致设施断电。在任何“危险轨迹”的风险评估中,将关键设备包含进来是很重要的。

---

[1] exclusion zone 也可翻译为隔离带或隔离区。

3.3 Examples of Line of Fire Hazards

Line of Fire hazard examples are shown from Figure 3-1 to Figure 3-12.

Figure 3-1 Moving vehicles/heavy equipment

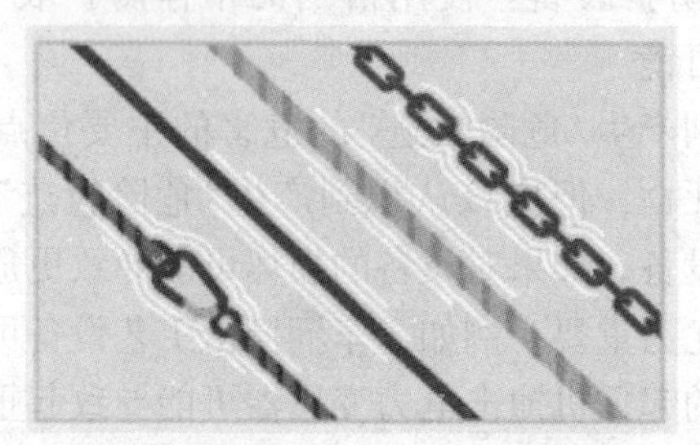

Figure 3-2 Objects under tension

Figure 3-3 Pressurized object

3.3 产生“危险轨迹”危害的示例

产生“危险轨迹”危害的示例如图 3-1～图 3-12 所示。

图 3-1　移动的机动车辆/重型设备

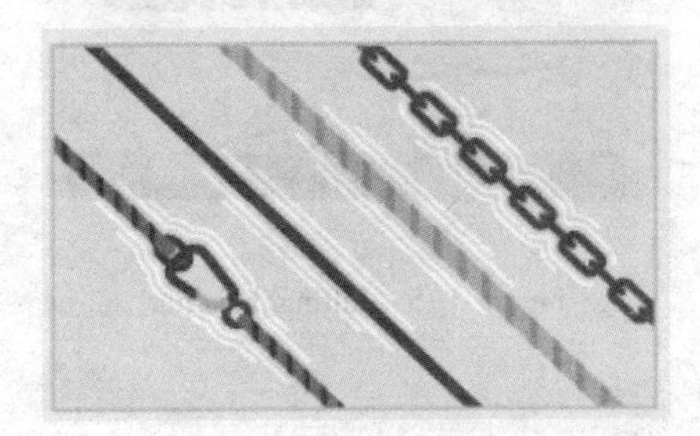

图 3-2　具有张力的物体

图 3-3　加压容器/管线

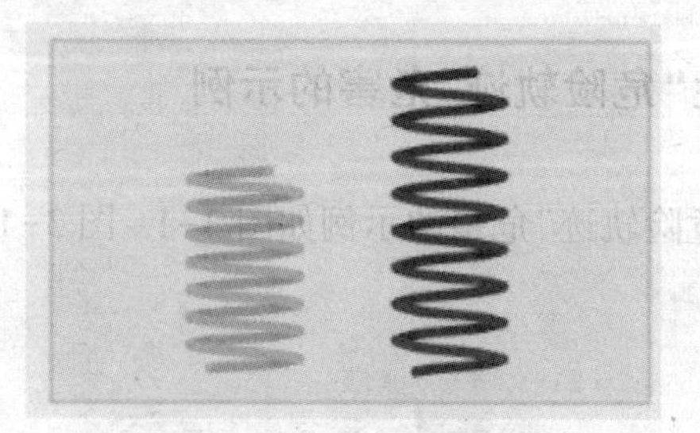

Figure 3-4 Spring-loaded devices

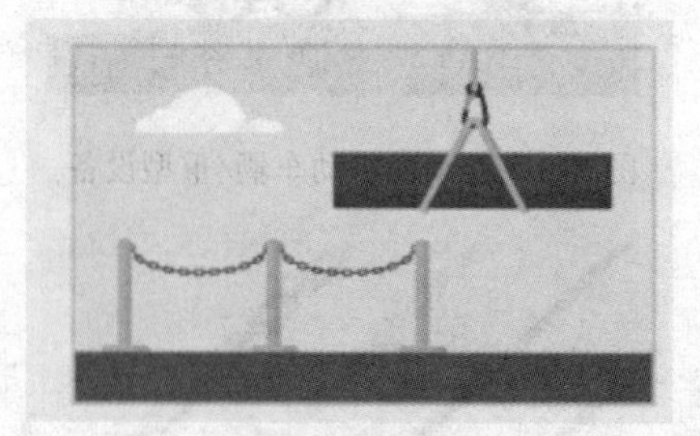

Figure 3-5 Lifting/hoisting

Figure 3-6 Objects with fall potential

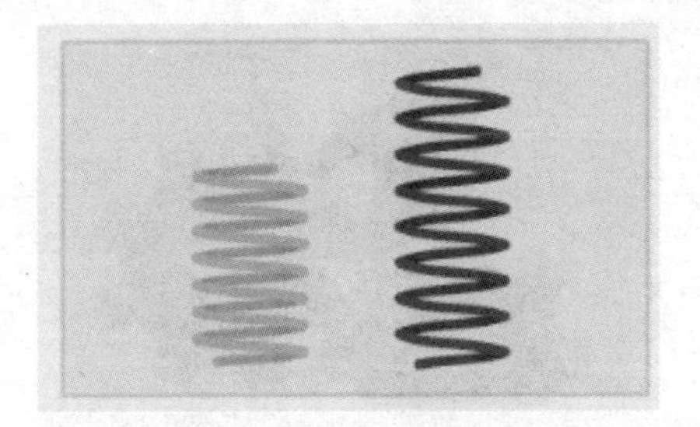

图 3-4 弹簧加载装置

图 3-5 吊装/起重

图 3-6 可能坠落的物体

Figure 3-7 Machinery with moving parts

Figure 3-8 Hand and power tools

Figure 3-9 Working at height/dropped objects

图 3-7 带有运动部件的机械

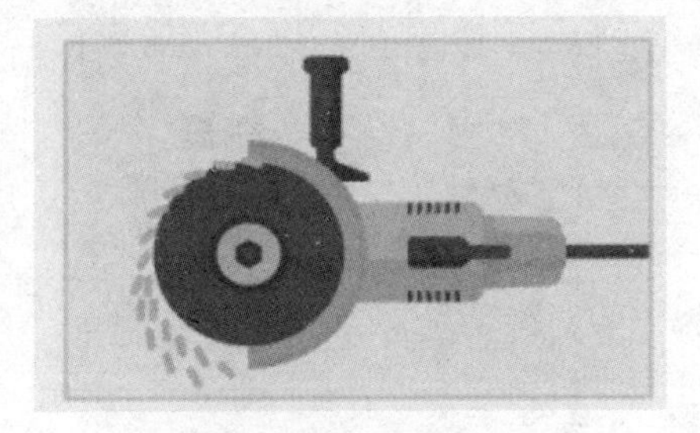

图 3-8 手工具和动力工具

图 3-9 高处作业/坠落物

Figure 3-10 Objects with roll potential

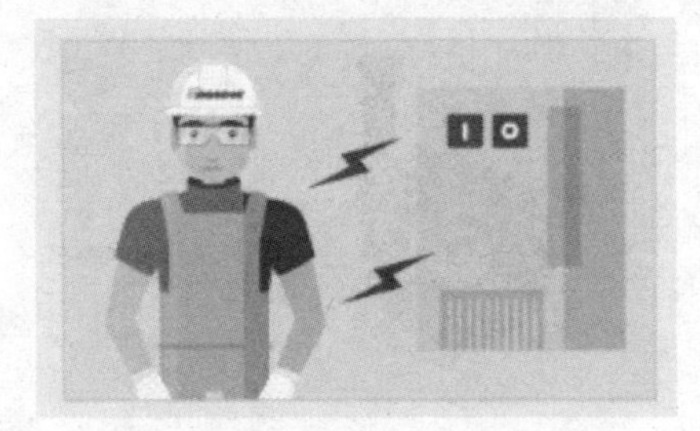

Figure 3-11 Electrical equipment

Figure 3-12 Ground disturbance and excavations

图 3-10 可能滚动的物体

图 3-11 电气设备

图 3-12 动土和开挖

3.4 Line of Fire Controls

There are a multitude of controls which can be used to eliminate or reduce the risk from line of fire hazards. These include:

- Lifting/hoisting controls:
  - *Never walk under a suspended load.*
  - *Is load balanced correctly prior to lifting?*
  - *Are exclusion zones in place?*
  - *Communicate to other workers when entering a lifting/hoisting zone, even if for a short period.*
  - *Has rigging gear been inspected?*
  - *Is rigging equipment used in excess of its maximum safe loading limit?*
  - *Have load swing hazards been considered?*
  - *Are tag lines used to ensure there is enough distance between you and the load?*
  - *Where would aperson fall to if the load they are pushing or pulling moves suddenly, or the surface they are standing on was to give way?*
- Moving vehicles/heavy equipment controls:
  - *Are parking braks and wheel chocks used?*
  - *Does operator have a full view of the working environment?*
  - *Are barricades/signs in place?*
  - *Can a person be located in between a piece of mobile equipment and another object?*
- Hand and power tools controls:
  - *Are adequate barriers and/or guards in place in case of kickback or unexpected release?*

3.4 “危险轨迹”控制措施

有许多控制措施可用于消除或降低来自“危险轨迹”危害的风险。这些包括：

- 吊装/起重控制措施：
  - 切勿在悬吊载荷下行走。
  - 吊装前(保留的)载荷余量是否正确?
  - 吊装作业是否设立了隔离带?
  - 进入吊装/起重区时，与吊装/起重工作人员进行沟通，即使仅仅短时间进入上述区域。
  - 是否检查过索具?
  - 是否超过所使用起重设备的最大安全载荷?
  - 是否考虑过载荷摆动的危险?
  - 牵引绳或溜绳是否用于确保您和载荷之间有足够的距离?
  - 如果工作人员推或拉的载荷突然移动，或者站立的表面坍塌，他会摔到哪里?
- 机动车辆/重型设备相关控制措施：
  - 是否使用了停车制动器和车轮垫块?
  - 操作人员是否全面了解工作环境?
  - 路障/标志是否到位?
  - 人是否可以位于移动设备和另一个物体之间?
- 手工具和动力工具控制措施：
  - 如果出现反冲(后坐)或意外漏电/漏气/漏液，足够的屏障和/或防护是否到位?

- *When using knives or cutting tools, cut away from your body.*
- *Are flying debris hazards considered? (grinding, welding, chiseling, etc.)*
- *Could the tool, or the object being worked on, impact a person if it was to slip?*

- Moving parts controls:
  - *Could equipment rotate unexpectedly?*
  - *Is guarding in place?*
- Electrical equipment control:
  - *Are workers outside the arc flash boundary when conducting switching operations?*
  - *Has the equipment been isolated and lock out/tag out applied?*
- Objects with roll potential control:
  - *Are adequate bracing/controls in place to prevent objects rolling unexpectedly?*
- Objects with fall potential controls:
  - *Are exclusion zones established under work performed at heights?*
  - *Are tools secured, kickboards fitted to working platforms?*
  - *Is mobile plant top heavy when in use, such as moving large loads at height with a forklift?*
  - *Are spoil piles located away from the edges of excavations?*
- Tensioned lines control:
  - *Are all personnel clear of tensioned cables, straps, chains and ropes?*
  - *Are correct gripping devices used?*
  - *Is selection of proper equipment based on size and load limit?*

- 当使用刀子或切割工具时，向远离身体(方向)进行切割。
- 是否考虑(磨、焊、凿等)飞溅碎屑危害?
- 当工具或被加工的物品滑脱时，是否会对人产生影响?

• 运动部件控制措施:
  - 设备是否会意外旋转?
  - 防护罩是否到位?
• 电气设备控制措施:
  - 进行开关作业时，工作人员是否处在电弧闪光边界之外?
  - 设备是否已隔离并上锁/挂签?
• 可能滚动物体的控制措施:
  - 是否有足够支撑/限制以防止物体意外滚动?
• 可能坠落物体的控制措施:
  - 在高处作业下方是否设立了隔离带?
  - (高处作业的)工具是否良好保护? 工作平台是否安装了踢脚板?
  - 移动设备在使用时是否顶部很重? 例如用叉车举升并移动重的载荷。
  - 挖掘出的土石方是否远离挖掘作业面边缘?
• 拉伸物体(电缆、皮带、链条等)的控制措施:
  - 所有人员是否远离处于拉伸状态的电缆、皮带、链条和绳索?
  - 是否使用了正确的夹具?
  - 是否根据尺寸和载荷限制选择合适的设备?

- *Are spring loaded devices opened slowly (actuators, etc.)?*
- *Has tension dissipated over time or does the energy still exist?*

• Pressurised systems controls:
- *Has the equipment been isolated and lock out/tag out applied?*
- *Has pressure been released to a safe location?*

3.5 Related Life-Saving Rules

Life-Saving Rule 3: Do not walk under a suspended load

Life-Saving Rule 9: Prevent dropped objects

Life-Saving Rule 10: Position yourself in a safe zone in relation to moving and energised equipment

Life-Saving Rule (New): Keep yourself and others out of the line of fire

3.6 References

(1) International Association of Oil & Gas Producers (IOGP), Life-Saving Rules https://www.iogp.org/life-savingrules/.

(2) UK HSE, Fatal injuries in Great Britain, 2017/18. http://www.hse.gov.uk/statistics/fatals.htm.

- 是否缓慢打开弹簧加载装置(执行机构等)?
- 张力随时间推移是否消散或能量是否仍然存在?

• 加压系统控制措施:
- 设备是否已隔离并上锁/挂签?
- 压力是否已释放到安全位置?

3.5 相关保命法则

保命法则 3：禁止从吊物下穿行

保命法则 9：防范坠落物

保命法则 10：时刻跟机动设备保持安全距离

保命法则（新）：让自己及他人远离危险的波及范围

3.6 参考资料

(1) International Association of Oil & Gas Producers (IOGP), Life-Saving Rules https://www.iogp.org/life-savingrules/.

(2) UK HSE, Fatal injuries in Great Britain, 2017/18. http://www.hse.gov.uk/statistics/fatals.htm.

4

BOWTIE ANALYSIS

领结图分析法

4.1 Introduction

The Oil & Gas industry is considered as a dangerous, or hazardous, industry. It is important that we protect our people and our business (asset, products, production capacity, reputation, etc.) from dangerous/hazardous occurrences. The hazards we encounter are the day-to-day hazards as part of our work (slips, trips, etc.) and those that can cause extreme or catastrophic consequences (fires, explosions, well blow-outs, etc.). These large events are commonly known as Major Accidents, and the hazards that can create them are known as Major Accident Hazards (MAHs).

There is legislation in many countries which require companies with the potential for MAHs to identify and properly manage those MAHs associated risks [such as the UK Control of Major Accident Hazards Regulations 2015, and the Workplace Safety and Health (Major Hazard Installations) Regulations 2017 from Singapore]. There are also a number of internationally recognised standards as well as a vast array of guidance documents related to MAH management.

The bowtie diagram (Figure 4-1) is a popular means of analysing, visualising and communicating complex MAH risk management relationships. Bowties are believed to have been developed by ICI in the 1970s and were championed by Royal-Dutch Shell which was the first major company to fully integrate bowties into its business practices.

4.1 简介

石油和天然气行业被认为是危险的行业。我们想方设法保护我们的员工和我们的业务(资产、产品、生产能力、声誉等)免受各种危险/危害事件的影响，这一点是至关重要的。我们遇到的危害是作为我们工作的一部分的日常危害(滑倒、绊倒等)以及可能导致极端或灾难性后果的危害(火灾、爆炸、井喷等)。这些大型事件通常被称为重大事故，可能产生这些事件的危害称为重大事故危害(MAHs)。

许多国家都有立法，要求有潜在重大事故危害(MAHs)的公司确定并妥善管理重大事故危害(MAHs)相关风险[例如2015年英国的重大事故危害控制法和2017年新加坡的工作场所安全与健康(重大危险装置)法]。还有一些国际公认的标准以及与重大事故危害(MAH)管理相关的大量指南性文件。

领结图(图4-1)是分析、可视化和交流复杂的重大事故危害(MAH)风险管理关系的常用方法。领结图被认为是由帝国化学工业公司(Imperial Chemical Industries Ltd.，ICI)在20世纪70年代开发的，并得到了荷兰皇家壳牌公司的支持。壳牌公司是第一家将领结图分析方法完全融入其商业实践的大公司。

The diagram is shaped like a bowtie (hence the name), creating a clear differentiation between proactive and reactive risk management. The power of the bowtie diagram is that it gives an overview of multiple plausible scenarios, in a single picture. In short, it provides a simple, visual explanation of a risk that would be much more difficult to explain otherwise.

Bowtie diagrams illustrate the controls in place to prevent, mitigate and recover from an MAH event defined by the Hazard and Top Event pair. The diagram presents the Threats which can cause the Top Event and the Preventive Barriers (left hand side) and the Consequences of the event and the Reactive Barriers (right hand side).

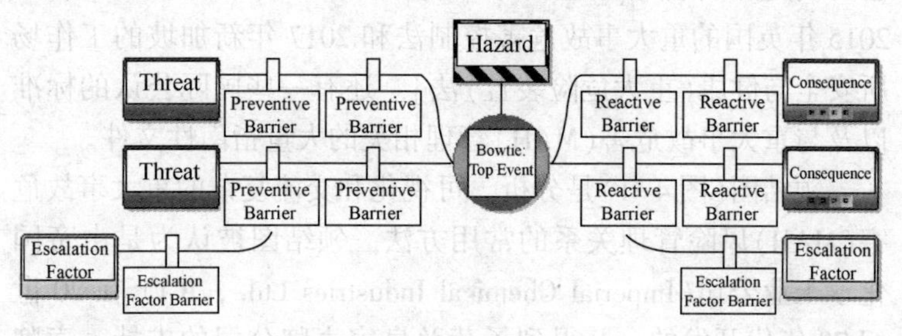

Figure 4-1 Bowtie Diagram

Hazard ——The potential to harm people and the environment, cause damage or loss of assets, and to adversely impact on reputation.

The start of any bowtie is the "hazard". A hazard is something in, around, or part of the organization which has the potential to cause damage. For instance, working with hazardous substances and driving a car are hazardous activities of an organization, while in general reading a book is not.

领结图的形状像一个领结(因此得名)，并在主动和被动风险管理之间设置了清晰的分界线。领结图的强大之处在于它在单个图表中给出了多个模拟合理情景的概览。简而言之，它提供了一种有关风险的简单、直观的解释，这是其他方法很难实现的。

领结图以图表的形式说明了用于防止、缓解和恢复由危害和顶上事件对定义的重大事故危害(MAH)相关事件的控制措施。该图显示了可能导致顶上事件的威胁和主动预防型屏障(左侧)以及事件后果和被动反应型屏障(右侧)。

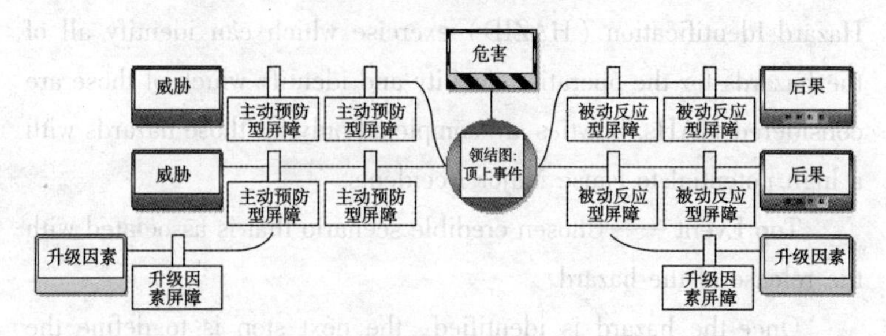

图 4-1 领结图

危害[1]——可能损害人和环境，造成财产损坏或损失，并对声誉产生不利影响。

任何领结图的开始都是“危害”。危害可能造成损害或损坏，它们是在组织内部、组织周围的某种事物，或者是组织的一部分。例如，使用有害物质和驾驶汽车是组织的危险活动，而阅读一本书通常不是(组织的危险活动)。

---

[1] 根据上下文，hazard 还可以翻译为危险、危险源、危险有害因素等。

The idea behind identifying hazards as part of a bowtie, is to find the things that are part of your organization that could have a negative impact if control over some aspects of that part are lost. They should be normal aspects of the operation/facility and not singular events or hazards that only happen in very extreme circumstances (these can be assessed separately). The rest of the bowtie is devoted to how we keep that normal but hazardous aspect from turning into something unwanted.

It is common to start the bowtie process by completing a Hazard Identification (HAZID) exercise which can identify all of the hazards for the operation/facility and identify which of those are considered MAHs. Bowties are completed only for those hazards with a high potential to cause Major Accidents.

Top Event ——Chosen credible scenario that is associated with the release of the hazard.

Once the hazard is identified, the next step is to define the "top event". This is the moment when control is lost over the hazard. There is no damage or negative impact yet, but it is imminent. This means that the top event is chosen just before events start causing actual damage (Figure 4-2).

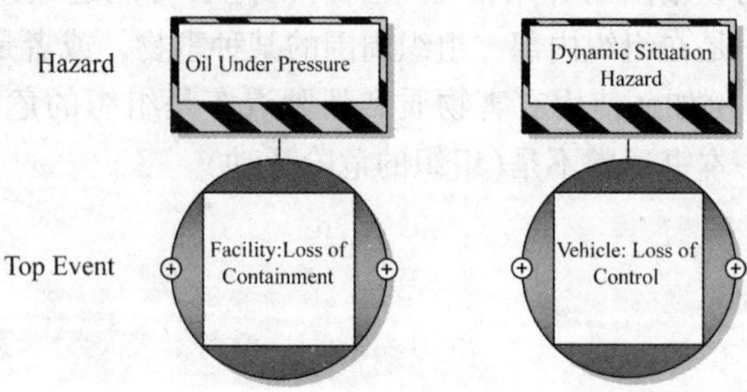

Figure 4-2 Hazard and Top Event Pair

将识别出的危害作为领结图一部分的想法是，找到组织的某个部分，如果失去对该部分的某个方面的控制，则可能产生负面影响[1]。它们应该是操作/设施的正常方面，而不是仅在极端情况下发生的不常见的事件或危害(这些可以进行单独评估)。领结图其余部分致力于我们如何防止正常但有危险的方面变成不需要的东西。

通常在完成危害辨识(HAZID)过程的基础上，构建领结图。该过程可以识别操作/设施的所有危害，并识别哪些危害被认为是重大事故危害(MAHs)。领结图仅针对那些可能导致重大事故的危害。

顶上事件——选择的与危害释放相关的可靠情景。

一旦识别出危害，下一步就是定义“顶上事件”。这是对危害失去控制的时刻。此刻尚未造成任何损害或负面影响，但是即将发生。这意味着，在事件刚要开始导致实际破坏之前，选定顶上事件(图 4-2)。

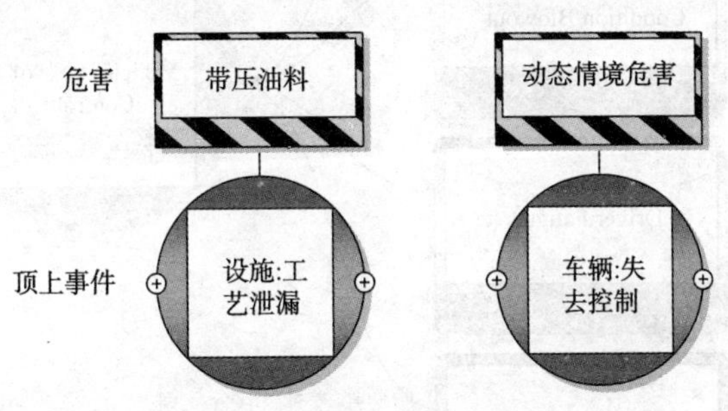

图 4-2　危害与顶上事件对

[1]“带压油料”是工艺过程的正常组成部分。但是，失去对“带压油料”某方面(温度、压力、流量等)的控制，就可能产生负面影响。

Threat ——A possible cause that will potentially release a hazard and produce a Top Event.

"Threats" are whatever will cause the top event (Figure 4-3). There can be multiple threats. Generic formulations like "human error", "equipment failure" or "weather conditions" should be avoided unless they have a direct impact on the top event. What does a person actually do to cause the top event? What kind of weather or what does the weather impact?

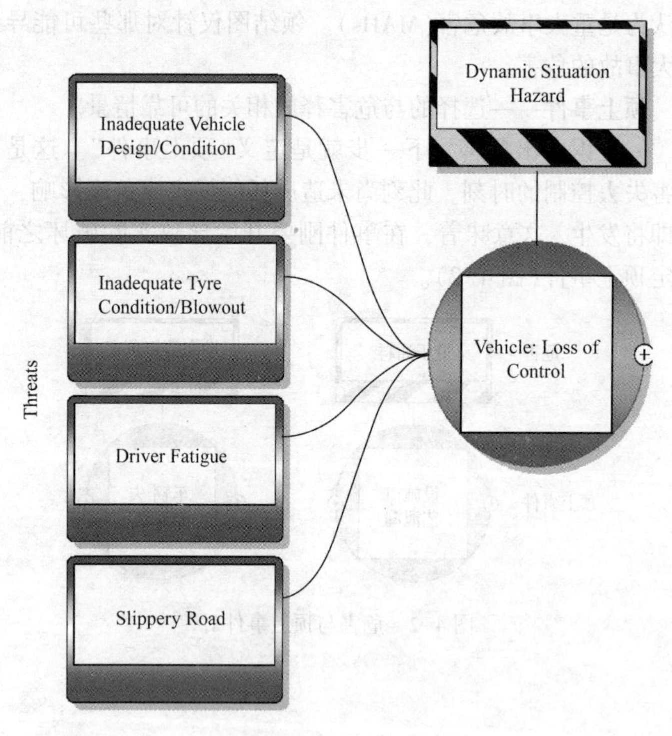

Figure 4-3 Threats

威胁——导致危害释放、引起顶上事件发生的可能原因。

“威胁”是导致顶上事件的任何因素(图4-3)。同一顶上事件，可能存在多种威胁。应避免使用“人为错误”“设备故障”或“天气状况”等通用表述，除非它们对顶上事件有直接影响。一个人实际上做了什么来引发顶上事件？什么样的天气或天气会影响什么？

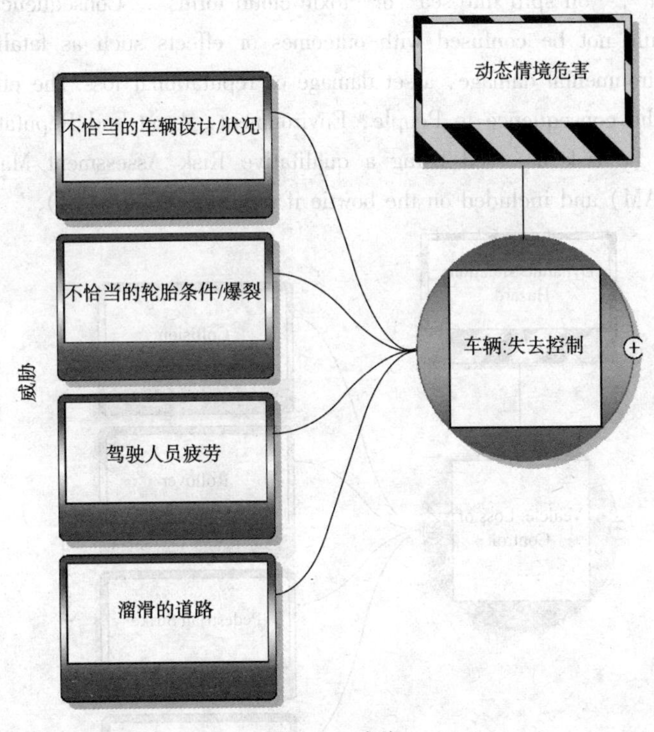

图4-3　威胁

Consequences ——An event or chain of events that result from the release of a hazard (generally leading to some form of harm).

"Consequences" are from the top event. There can be more than one consequence for every top event. As with the threats, people tend to focus on generic categories instead of describing specific events. "Consequences" should be described as actions such as "car roll over", "oil spill into sea" or "toxic cloud forms". "Consequences" should not be confused with outcomes or effects such as fatality, environmental damage, asset damage or reputational loss. The effect of the consequence to People, Environment, Asset and Reputation can be risk assessed using a qualitative Risk Assessment Matrix (RAM) and included on the bowtie if required(Figure 4-4).

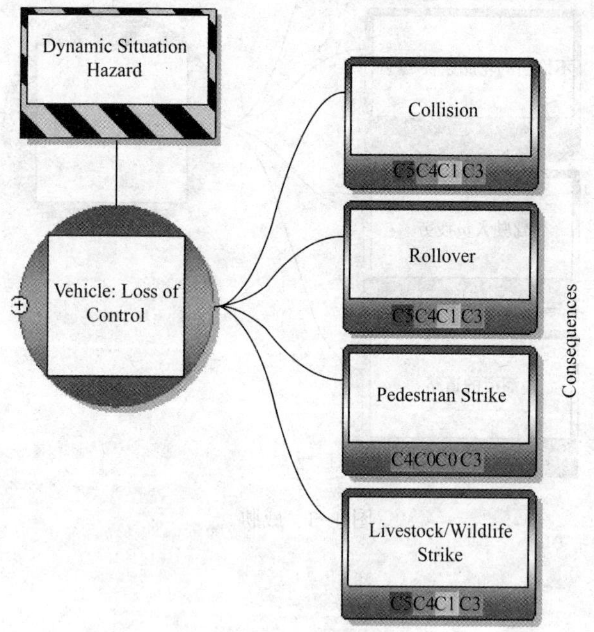

Figure 4-4 Consequences

后果——由于危害释放而导致的事件或事件链(通常会导致某种形式的损害)。

"后果"来自顶上事件。每个顶上事件都有不止一个后果。与威胁一样，人们倾向于关注通用类别而不是描述特定事件。"后果"应该被描述为"汽车翻车""原油泄漏进入大海"或"形成有毒云状物"等基本事件。不应将"后果"与死亡、环境破坏、资产损坏或声誉损失等结果或影响相混淆。"后果"对人员、环境、资产和声誉的影响，可以使用定性风险评估矩阵(RAM)进行风险评估，并在必要时包含在领结图中(图 4-4)。

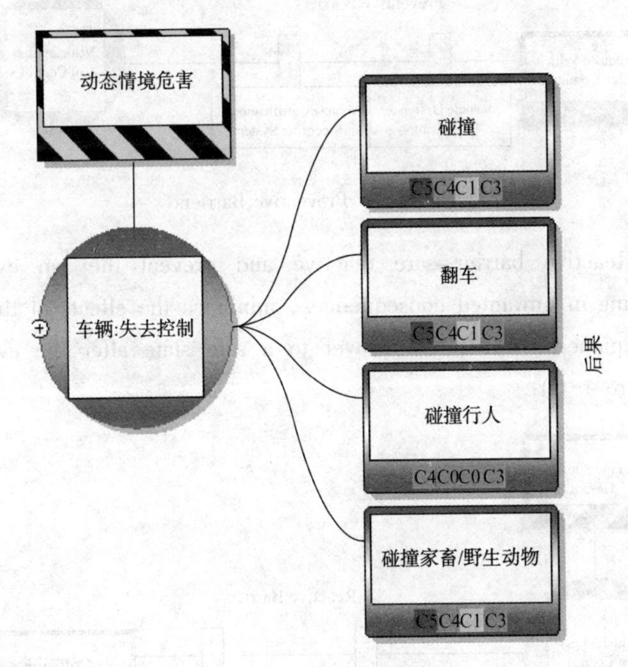

图 4-4 后果

Barriers —Any measure taken that acts against some undesirable force or intention, in order to maintain a desired state.

Barriers in the bowtie appear on both sides of the top event. Preventive barriers interrupt the scenario so that the threats do not result in the top event (Figure 4–5).

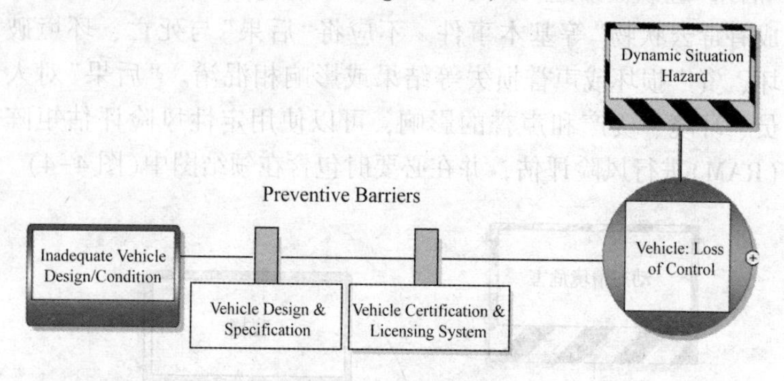

Figure 4–5 Preventive Barriers

Reactive barriers are reactive and prevent the top event resulting in unwanted consequences, minimise the effects of those consequences or help us recover to a safe state after the event (Figure 4–6).

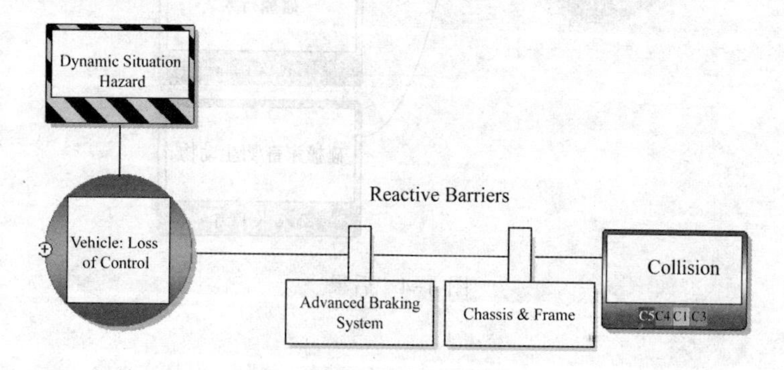

Figure 4–6 Reactive Barriers

屏障——针对某些不期望的力量或意图采取的任何措施，以实现或维持期望的状态。

领结图上的屏障出现在顶上事件的两侧。主动预防型屏障会终止某种情景，以使威胁不会导致顶上事件(图 4-5)。

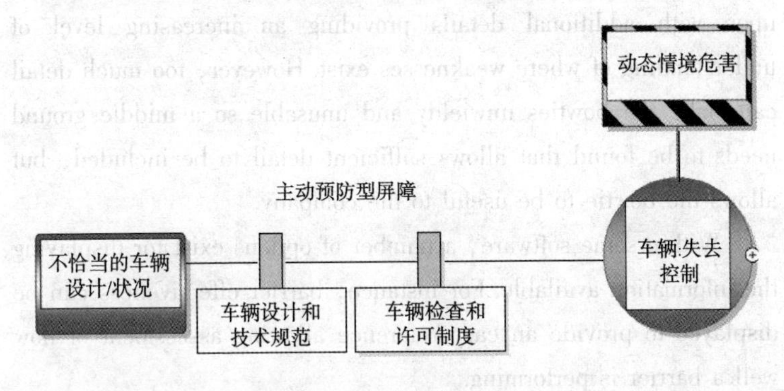

图 4-5 主动预防型屏障

被动反应型屏障是被动性的，可防止顶上事件导致不需要的后果，最大限度地减少这些后果的影响或帮助我们在事件发生后恢复到安全状态(图 4-6)。

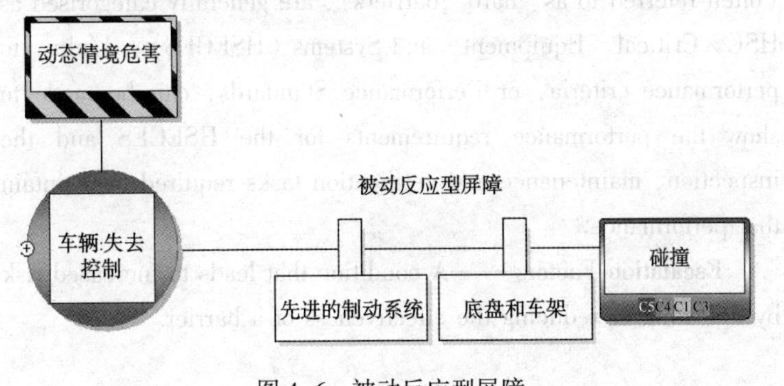

图 4-6 被动反应型屏障

There are different types of barriers, which are mainly a combination of human behaviour and/or hardware/technology. Once the barriers are identified, a basic understanding of how risks are managed is established. This basic barrier structure can be built upon with additional details providing an increasing level of understanding of where weaknesses exist. However, too much detail can make the bowties unwieldy and unusable so a middle ground needs to be found that allows sufficient detail to be included, but allows the bowtie to be useful to the company.

Within some software, a number of options exist for displaying the information available. For instance, barrier effectiveness can be displayed to provide an easy reference allowing assessment of how well a barrier is performing.

For barriers which rely on human behaviour (often referred to as "soft" barriers), the Safety Critical Tasks (SCTs) required to implement and maintain the barrier, can be presented along with the responsible party for the SCT. Hardware/technology barriers (often referred to as "hard" barriers), are generally categorised as HSE Critical Equipment and Systems (HSECES). Links to performance criteria, or Performance Standards, can be made to show the performance requirements for the HSECES and the inspection, maintenance and verification tasks required to maintain that performance.

Escalation Factors ——A condition that leads to increased risk by defeating or reducing the effectiveness of a barrier.

存在不同类型的屏障，它们主要是人的行为和/或硬件/技术的组合。一旦确定了屏障，就建立了对风险管理方式的基本了解。可以在基本屏障结构基础上补充更多细节，提供对存在薄弱环节的地方越来越多的理解。然而，过多的细节可能会导致领结图庞杂且无法使用，因此需要找到允许包含足够细节的中间地带，并使领结图对企业有用。

在某些软件中，存在许多用于显示可用信息的选项。例如，可显示屏障有效性以提供简单的参考信息，辅助屏障性能的评估。

对于依赖人的行为的屏障(通常称为“软”屏障)，实施和维护屏障所需的安全关键任务(SCTs)可以与其责任方一起提出。硬件/技术屏障(通常称为“硬”屏障)通常被归类为 HSE 关键设备和系统(HSECES)。可以设置 HSECES 性能准则或性能标准的链接，以显示 HSECES 的性能要求以及维持该性能所需的检查、维护和验证任务。

升级因素——通过破坏(打垮)屏障或降低屏障有效性，导致风险增加的条件。

Barriers are never perfect. Even the best hardware barrier can degrade over time and fail. Given this fact, the reasons why the barrier could fail needs to be understood. This is done using a type of threat box known as an "escalation factor". Anything that will make a barrier fail can be described in an escalation factor.

Care should be taken not to include escalation factors that are just the opposite of the barrier being assessed. For example, if the barrier is "maintenance" the escalation factor should not be "maintenance not performed". This is just a failure of the "maintenance" barrier and not an escalation factor. An escalation factor in this case may be "maintenance not performed due to lack of spares".

Escalation Factor Barriers —Barriers put in place to manage conditions that lead to increased risk due to loss of Preventive or Reactive Barriers.

The escalation factor barriers for the example "maintenance not performed due to lack of spares" could be "critical spares list" and "automatic ordering of critical spares" (Figure 4-7).

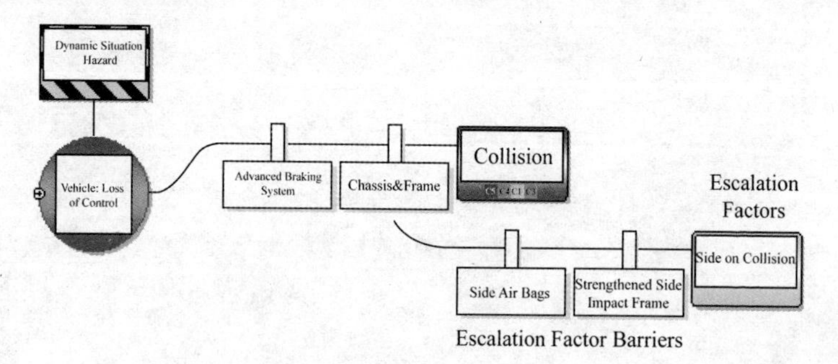

Figure 4-7 Escalation Factors

从来没有完美的屏障。即使是最好的硬件屏障也会随着时间的推移而降级并失效。鉴于这一事实，需要了解屏障可能失效的原因。这是通过使用一种被称为“升级因素”的威胁框来实现的。任何使屏障失效的因素都可以在“升级因素”中描述。

应注意不要包括与待评估屏障内容正好相反的升级因素。例如，如果屏障是“维护”，则升级因素不应为“不执行维护”。这只是“维护”屏障的失效，而不是升级因素。在这种情况下，升级因素可能是“由于缺少备件而未执行维护”。

升级因素屏障——为管理导致主动预防型屏障或被动反应型屏障失效并使风险增加的状况（因素）而设置的屏障。

例如“由于缺少备件而未执行维护”的升级因素屏障可能是“关键备件清单”和“自动订购关键备件”（图4-7）。

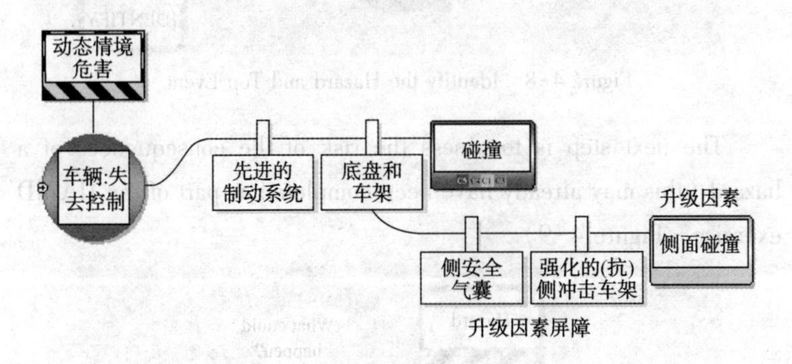

图4-7　升级因素

4.2 Bowtie Development

The first activity is to identify the hazards and the associated top event(Figure 4-8).

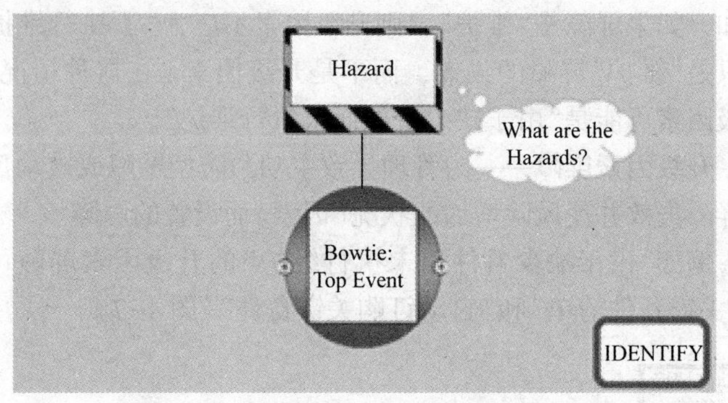

Figure 4-8 Identify the Hazard and Top Event

The next step is to assess the risk of the consequences of a hazard (this may already have been completed as part of the HAZID exercise, Figure 4-9).

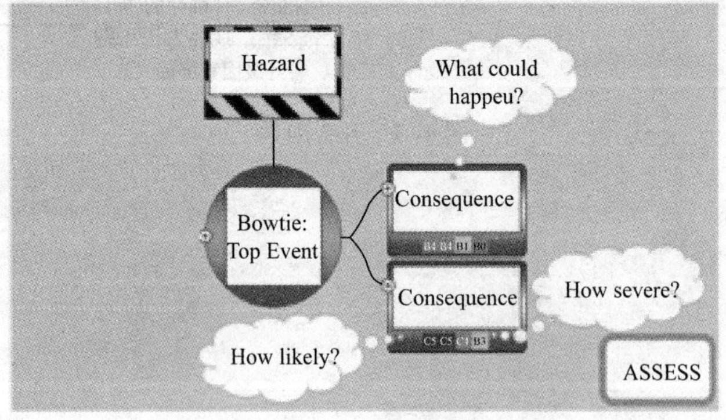

Figure 4-9 Assess the Consequences

4.2 领结图的编制

领结图分析的第一步是确定危害及其关联顶上事件(图 4-8)。

图 4-8 识别危害及其关联顶上事件

下一步是评估危害后果的风险(这可能已经作为 HAZID 过程的一部分完成了, 图 4-9)。

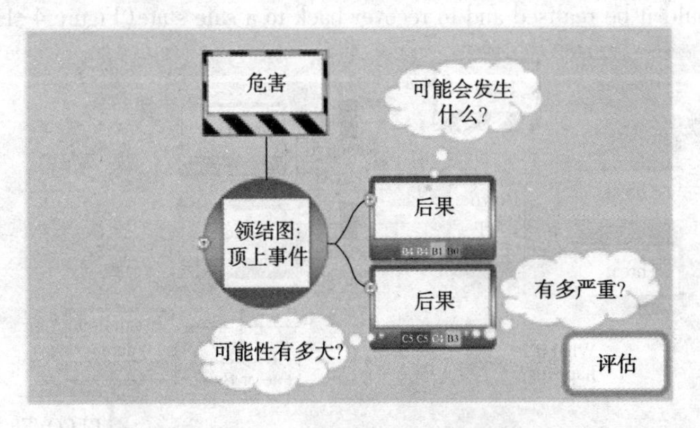

图 4-9 评估危害后果的风险

After we have assessed the risks, we can identify the barriers we have in place to prevent the top event from being realised, or ways we can control the hazard (Figure 4-10).

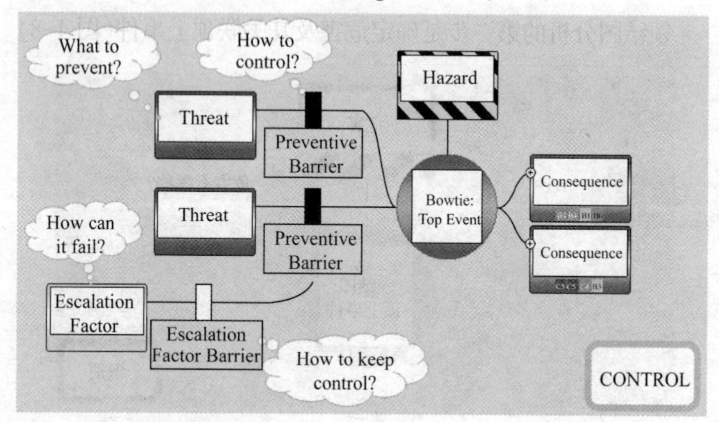

Figure 4-10 Identify the Threats and the Preventive Controls

As stated earlier, no barrier is 100% effective and so we have to ensure that there are ways to reduce the effects of the top event should it be realised and to recover back to a safe state (Figure 4-11).

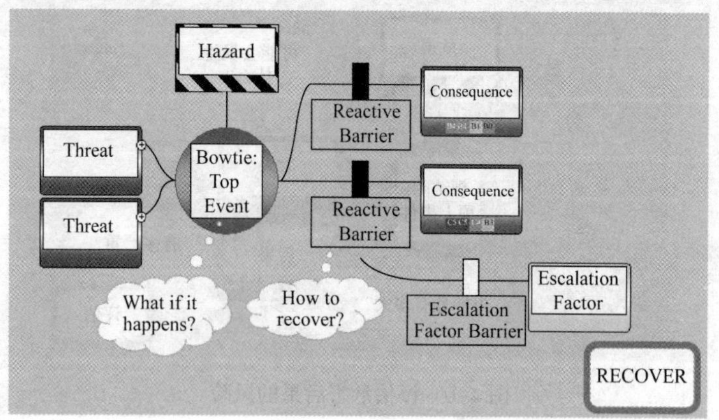

Figure 4-11 Identify the Reactive Controls

在我们评估了风险之后，我们可以确定我们已经设置的防止顶上事件发生的屏障，或者我们可以控制危害的方法(图4-10)。

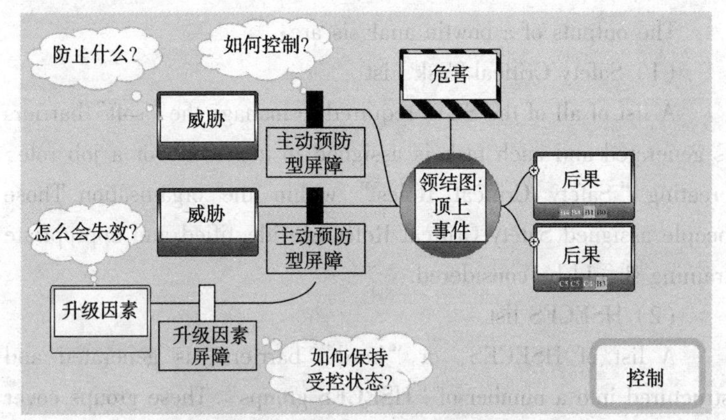

图4-10 确定威胁和主动预防型措施

如前所述，没有屏障是100%有效的。因此，如果顶上事件发生了，我们必须确保有一些方法可以减少其影响，并恢复到安全状态(图4-11)。

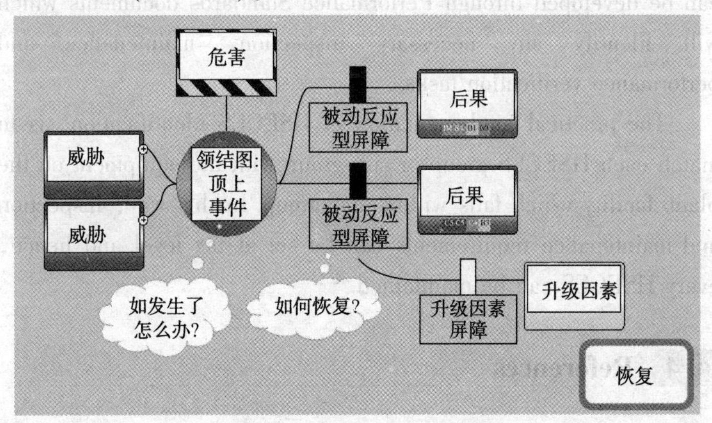

图4-11 确定被动反应型措施

4.3 Outputs from Bowtie Analysis

The outputs of a bowtie analysis are:

(1) Safety Critical Task List

A list of all of the SCTs required to manage the "soft" barriers is generated and each task is assigned to a person, or a job role, creating "Safety Critical Roles" within the organisation. Those people assigned Safety Critical Roles are identified and appropriate training should be considered.

(2) HSECES list

A list of HSECES, or "hard" barriers, is generated and structured into a number of "HSECES groups". These groups cover topics such as structural integrity, process containment, detection, shutdown, etc. The bowtie will normally identify the HSECES group (and sub-groups as necessary) that the barrier belongs to. Performance requirements for each HSECES group or sub－group can be developed through Performance Standards documents which will identify any necessary inspection, maintenance and performance verification tasks.

The practical implementation of HSECES identification, is to match each HSECES group or sub-group with the equipment on the plant/facility which falls within that group. In this way, inspection and maintenance requirements can be set at tag level and hence, every HSECES can be maintained.

4.4 References

CGE Risk, The Bowtie Method.

4.3 领结图分析的输出

领结图分析的输出是：

(1) 安全关键任务(SCT)清单

生成管理“软”屏障所需的所有安全关键任务(SCTs)清单，并将每项任务分配给一个人或一个工作角色，从而在组织内创建“安全关键角色”。确定那些承担安全关键角色的人员，并应考虑对其进行适当培训。

(2) HSE关键设备和系统(HSECES)清单

生成HSE关键设备和系统(HSECES)或“硬”屏障清单，并将其组织成若干“HSE关键设备和系统(HSECES)组”。这些组涵盖的主题包括结构完整性、工艺防泄漏、探测、停车等。领结图通常会确定屏障所属的HSE关键设备和系统(HSECES)组(以及必要的子组)。每个HSE关键设备和系统(HSECES)组或子组的性能要求，可以通过性能标准文档进行开发，该文档将确定任何必要的检查、维护和性能验证任务。

确定HSE关键设备和系统(HSECES)的实用做法是将每个HSE关键设备和系统(HSECES)组或子组与工厂/设施中属于该组或子组的设备进行匹配。通过这种方式，可以在标签级别(设备)设置检查和维护要求，进而可以使每个HSE关键设备和系统(HSECES)得到良好维护。

4.4 参考资料

CGE Risk, The Bowtie Method.

5

BARRIER

屏 障

5.1 Introduction

There are a number of concepts with respect to Major Hazard Risk Management. One of them is Barrier Management. The general theme for this concept is to identify all of the barriers that are in place to either prevent a Major Accident, mitigate the consequences of a Major Accident or to recover from the consequences of a Major Accident and ensure they are effectively managed.

In the Oil & Gas Industry, it is a good practice to identify and manage barriers following the Bowtie methodology (other methodologies do exist). Hazards identified as high risk or very high consequence on the Risk Assessment Matrix (RAM) are individually assessed to determine how the hazard can be realised, the threats to (or causes of) the realisation and the consequences of that realisation. These threats and consequences are assessed to identify the controls in place to manage the hazard.

This methodology is also being widely used by other industries such as aviation, rail and healthcare, to name a few.

A barrier can be defined as any measure taken which acts against some undesirable force or intention, in order to achieve or maintain a desired state. In simple words, they are the controls that can be set in place to prevent things (generally negative things) from happening.

Barriers are often described as:

- Preventive Barriers——those barriers which prevent the hazardous event (hazard and top event pair) from occurring.

5.1 简介

关于重大危害风险管理有许多概念。其中一个概念是屏障管理。这一概念的核心主题是确定所有防止重大事故、减轻重大事故后果或从重大事故后果中恢复的屏障，并确保其得到有效管理。

在石油和天然气行业，根据领结图分析法(当然也存在其他方法)来识别和管理屏障是一种好的做法。对风险评估矩阵(RAM)中确定为高风险或非常高后果的危害进行单独评估，以确定潜在危害如何转变为现实危害、促成上述转变的威胁(或原因)以及现实危害(潜在危害失控之后)的可能后果。评估这些威胁和后果，以确定管理重大事故危害的控制措施。

领结图分析法也被其他行业广泛使用，例如航空、铁路和医疗保健等。

屏障可以定义为针对某些不期望的力量或意图采取的任何措施，以实现或维持期望的状态。简单来说，它们是我们为防止事物(通常是负面事物)发生而采取的控制措施。

屏障通常被描述为：

- 主动预防型屏障——防止发生危险事件(危害和顶上事件对)的屏障。

• Reactive Barriers——those barriers which mitigate the effects of a hazardous event, or ones that recover from the effects of a hazardous event back to a safe state.

The following diagram shows a Bowtie diagram and where the barriers fit into Figure 5-1.

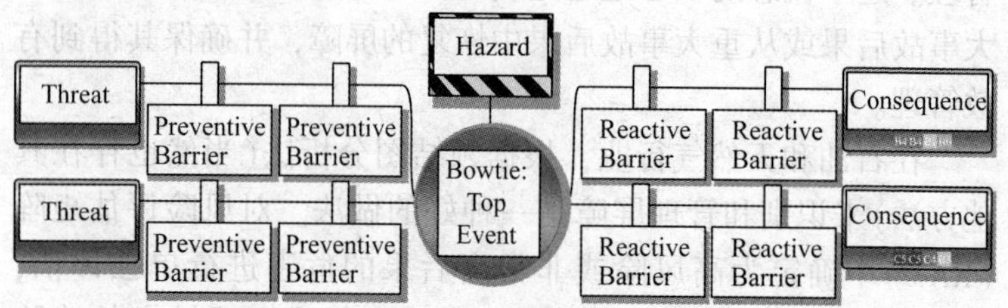

Figure 5-1 Bowtie Diagram

As can be seen from Figure 5-1, the barriers are represented as being directly in line with the threats and consequences. For each threat or consequence, the methodology assumes there will be multiple barriers, and in reality, this would be the case.

It is recognised that not all barriers are 100% effective and as such, there are generally numerous barriers in place. These barriers are often of different types and so the failure mechanisms for each of the barriers is different. The hope is that if one barrier fails in one way, this will not affect the other barriers and so the event will be prevented.

The bowtie methodology is based on the "Swiss Cheese Model of Accident Causation". This model was first described by James Reason in the 1980s showing that although barriers had "holes" in them (not fully effective) the more barriers that exist, the harder it is for all of the "holes" to line up so an accident (or hazardous event) can occur. The model can also be applied to the reactive side of the bowtie model showing that the more barriers that are in place, the less likely it is that the event will lead to the consequence.

• 被动反应型屏障——减轻危险事件影响的屏障，或从危险事件的影响恢复到安全状态的屏障。

如图 5-1 所示，显示了领结图以及屏障如何纳入领结图中。

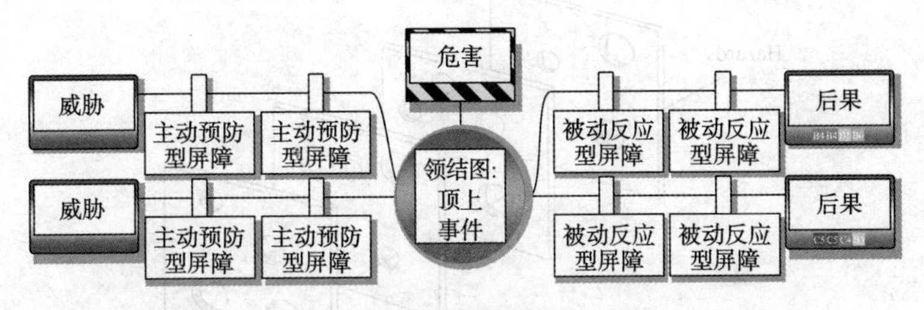

图 5-1　领结图

从图 5-1 中可以看出，屏障被表示为与威胁和后果直接对应。对于每种威胁或后果，该方法假设存在多个屏障，实际上情况就是如此。

人们认识到，并非所有屏障都是 100%有效的。因此，通常存在许多屏障。这些屏障通常是不同类型的，因此每个屏障的失效机制是不同的。期望的情况是，如果一个屏障以某种方式失效，不会影响其他屏障，从而将阻止危险事件的发生。

领结图分析法基于“事故致因的瑞士奶酪模型”。詹姆斯·理森在 20 世纪 80 年代首次描述了这种模型，虽然屏障中存在“漏洞”(并非完全有效)，但存在的屏障越多，所有“漏洞”排列在一起发生事故(或危险事件)的可能性就越小。该模型还可以应用于领结模型被动反应型屏障的一侧(右侧)，表明设置的屏障越多，事件导致后果的可能性就越小。

Figure 5-2 shows how the holes in the cheese line up to allow the hazardous event to occur. Each slice of cheese represents a barrier.

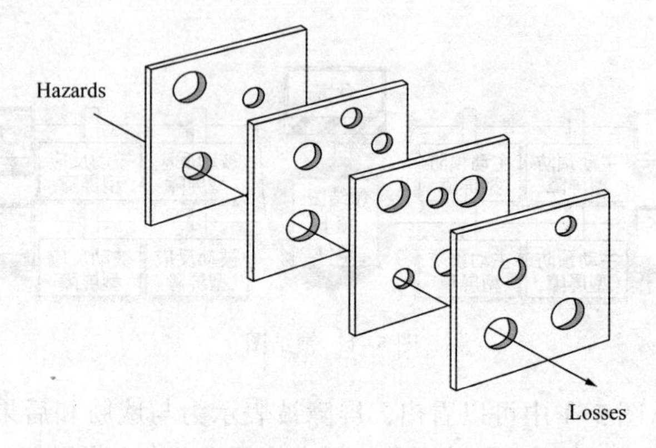

Figure 5-2 Swiss Cheese Model

5.2 Types of Barriers

Barriers can be further classified into two main types:

(1) Hardware Based Barriers

These are barriers made up of physical equipment such as pipes, vessels, pressure relief valves, shutdown systems, gas detectors, etc. This type of barrier will generally be described as a Safety Critical Element (SCE), or as HSE Critical Equipment and Systems (HSECES). These barriers can also be described as "Hard" barriers or hardware barriers.

图 5-2 显示了奶酪中的孔洞如何排列在一起导致危险事件的发生。每片奶酪代表一道屏障。

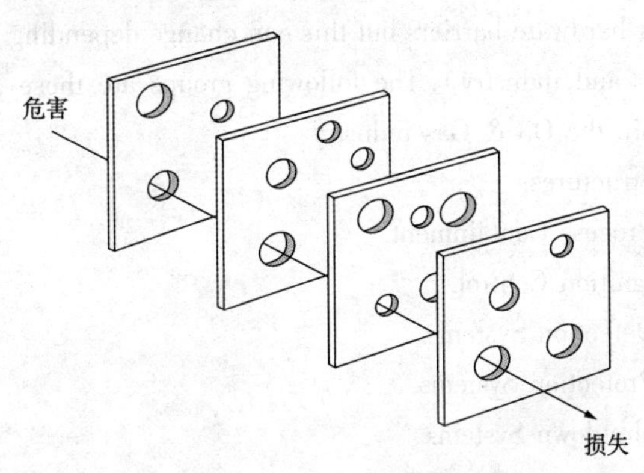

图 5-2 瑞士奶酪模型

5.2 屏障的类型

屏障可以进一步分为两种主要类型：

(1) 基于硬件的屏障

这些屏障由管道、容器、压力释放阀、停车系统、气体探测器等物理设备组成。这种类型的屏障通常被描述为安全关键部件(SCE)或 HSE 关键设备和系统(HSECES)。这些屏障也可以被称为“硬”屏障或硬件屏障。

As there are a large number of hardware barriers, it is normal for these to be grouped together. There are generally nine recognised groups for hardware barriers but this can change depending upon the company (and industry). The following groups are those generally used within the Oil & Gas industry:

① Structures.

② Process Containment.

③ Ignition Control.

④ Detection Systems.

⑤ Protection Systems.

⑥ Shutdown Systems.

⑦ Emergency Response.

⑧ Lifting Equipment.

⑨ Life-Saving Equipment.

As can be seen from the above list, HSECES barriers can be preventive and reactive. Figure 5-3 shows how these hardware barriers fit together to prevent the hazardous event from occurring (preventive barriers) and to mitigate or recover from the event (reactive barriers).

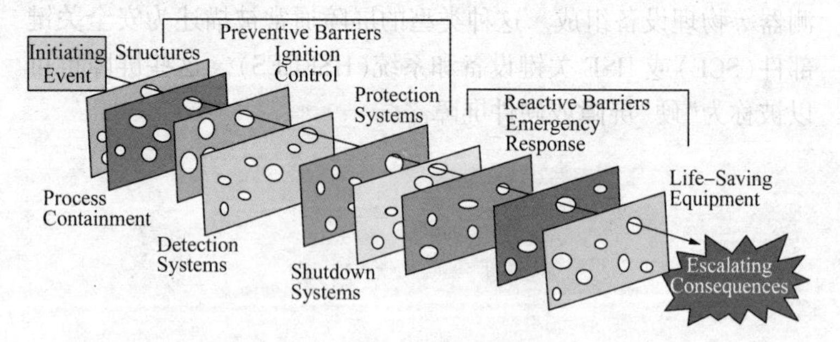

Figure 5-3 Hardware Barrier Groups

由于存在大量硬件屏障，因此将这些硬件屏障进行分组是正常的。通常有九个公认的硬件屏障分组，但分组方式可能会因公司(和行业)变化而变化。石油和天然气行业中常用以下分组方式：

① 结构；

② 工艺防泄漏；

③ 点火控制；

④ 探测系统；

⑤ 保护系统；

⑥ 停车系统；

⑦ 紧急响应；

⑧ 起重设备；

⑨ 救生设备。

从上面的分组中可以看出，HSE关键设备和系统(HSECES)屏障可以是预防型的和反应型的。图5-3显示了这些硬件屏障如何配合在一起以防止危险事件发生(主动预防型屏障)以及减轻事件后果或从事件中恢复(被动反应型屏障)。

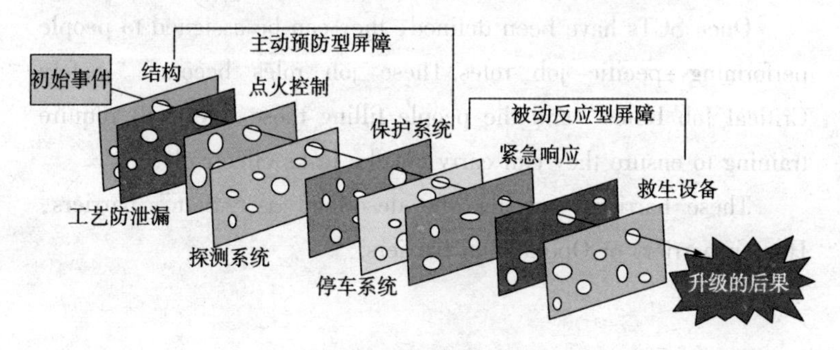

图5-3 硬件屏障组

Each of these groups is sub-divided into sub-groups that concentrate on specific types of controls. For example, Structures could be divided into below ground structures and above ground structures, etc. The sub-groups are likely to change depending upon the design of the facility.

(2) Task Based Barriers

This type of barriers require human actions for them to be effective. They may describe systems and procedures that are in place (such as PTW, SOPs, etc.) or may describe direct actions that must be taken (such as "operator shuts down vessel on high high alarm"). If the barrier relates to a system or a procedure, the tasks required are the development and implementation of the system/procedure and the effective use of the system/procedure.

The actions required for the operation of the system or procedure, or the direct action stated are known as Safety Critical Tasks (SCTs) or Safety Critical Activities (SCAs). These terms are generally interchangeable.

Once SCTs have been defined, they can be assigned to people performing specific job roles. These job roles become "Safety Critical Job Roles" and the people filling those roles will require training to ensure they can carry out the tasks effectively.

These barriers can also be described as "Soft" barriers, Human barriers or Operational barriers.

这些一级分组中的每一组被进一步分为专门针对特定类型控制措施的二级分组。例如，结构可以分为地下结构和地上结构等。二级分组可能会根据设施的设计而改变。

(2) 基于任务的屏障

这种类型的屏障要求人采取行动才能有效。它们可以描述现有的制度和程序[例如作业许可(PTW)、标准作业程序(SOPs)等]，或者可以描述必须采取的直接行动(例如“操作员在高-高警报时关闭容器”)。如果屏障跟一项制度或一个程序有关，则所需的任务是制度/程序的开发和实施，以及制度/程序的有效使用。

制度或程序运行所需的行动或者规定的直接行动，称为安全关键任务(SCTs)或安全关键活动(SCAs)。这些术语通常是可以互换的。

一旦定义了安全关键任务(SCTs)，就可以将它们分配给执行特定工作角色的人员。这些工作角色成为“安全关键工作角色”，填补这些角色的人员将需要培训，以确保他们能够有效地执行任务。

这些屏障也可以被描述为“软”屏障、行为屏障或操作屏障。

5.3 Performance Standards of Barriers

(1) Performance Standards——Hard Barriers

Hardware based barriers are expected to perform in a specific manner when they are called upon to do so. For HSECES barriers this means that each piece of equipment, will operate at its design intent, will be available when it is required to operate and will operate reliably. In addition, some equipment is required to operate after a major accident has occurred and as such, survivability criteria need to be defined where relevant.

The functionality, availability, reliability and survivability of an HSECES is generally detailed in a Performance Standard. The information from the performance standard is used to determine the maintenance and inspection schedule for the equipment. Assurance tasks help the company to be confident that the equipment is working and is performing its "hazard control" function effectively. This in turn, allows the company to determine if the HSECES barrier on the bowtie is effective or not.

(2) Performance Standards——Soft Barriers

Establishing the criteria that hard barriers need to meet is easier than establishing similar criteria for soft barriers. Machinery responds much more consistently than humans.

A task-based barrier Performance Standard is required to describe what will be completed, by whom and by when. The task-based barrier Performance Standard should also show reliability of the actions taken by identifying maximum acceptable number or percentage of missed events and/or false alarms.

5.3 屏障的性能/绩效标准

(1) 性能标准——硬屏障

当要求硬件屏障以特定方式发挥作用时，它们应满足相关要求。对HSE关键设备和系统(HSECES)类的屏障来说，这意味着每件设备都将按照其设计意图运行，在需要它们发挥作用时可以可靠运行。此外，一些设备需要在发生重大事故后运行。因此需要确定此类设备的耐受标准。

HSE关键设备和系统(HSECES)的功能性、可用性、可靠性和耐受性，通常在性能标准中详细说明。性能标准中的信息用于确定设备的维护和检查计划。保证任务有助于企业确信设备正在运行并有效地执行其“危险控制”功能，这随之允许企业确定领结图上的HSE关键设备和系统(HSECES)类的屏障是否有效。

(2) 绩效标准——软屏障

确定硬屏障需要满足的标准比建立类似的软屏障标准更容易。机器响应的一致性比人响应的一致性更强。

基于任务的屏障的绩效标准要描述什么任务内容要完成、由谁完成、何时完成。基于任务的屏障的绩效标准，也应该通过确定错失事件和/或误报警(虚警)的最大可接受数量或百分比，来说明所采取行动的可靠性。

For task-based barriers, the associated assurance tasks focus on inspections during the task being undertaken and audits of the system. Audit results allow the company to be confident that the system or procedure is being effectively implemented and that the "hazard control" function is effective. This in turn, allows the company to determine if the task-based barrier on the bowtie is effective or not.

5.4 Required Number of Barriers

As previously stated, not all barriers are 100% effective, in reality, there are very few barriers that can be considered 100% effective. Even very robustly designed mechanical systems, with computer aided logic can fail to operate when required.

It is important to ensure that the threat or the consequences are effectively controlled. This may be achieved by one or two barriers, or may require eight or ten or more barriers. The goal is to identify sufficient barriers for the effective control of the threat or consequence.

Care should be taken not to "double count" dependant barriers. For example, a level switch on a tank and an automatic stop on a pump may be considered as barriers to prevent overfilling of the tank. However, the pump will not automatically stop unless the level switch tells it to. Therefore, this is really only one barrier with two failure modes (failure of the switch and failure of the automatic stop signal).

对基于任务的屏障，相关保证任务侧重于在执行任务期间的检查和对制度体系的审核。审核结果使公司能够确信制度或程序得到有效实施，并且“危险控制”功能是有效的，这随之允许公司确定领结图上基于任务的屏障是否有效。

5.4 所需的屏障数量

如前所述，并非所有屏障都是100%有效。实际上，很少有屏障可以被认为是100%有效的。即使设计非常强健，带有计算机辅助逻辑的机械系统，也可能无法在需要时(正常)运行。

确保有效控制威胁或后果非常重要。这可以通过一个或两个屏障来实现，或者可能需要8个或10个或更多屏障来实现。目标是确定有效控制威胁或后果的足够屏障。

应注意不要“重复计算”相互依赖的屏障。例如，罐上的液位开关和泵上的自动停止功能可被视为防止溢罐的屏障。但是，除非液位开关发信号给泵，否则泵不会自动停止。因此，这实际上只是一个具有两种故障模式的屏障(开关故障和自动停止信号失效)。

Although some companies do prescribe a minimum number of barriers within their company standards, barrier counting can be counter-productive. Additional, and costly, barriers may be included that are not required in order to make up numbers, while on the other hand, too few barriers may be considered even if the required number is reached if they do not effectively control the hazard.

5.5 Traps and Tips of Barrier Identification

- **Trap:** Control descriptions that are too generic e. g. control: "Security Patrol".
- **Tip:** Describe what the control actually does with the reader in mind e. g.: "Security patrol identifies leak and informs control room". Try to include the action that takes place to interrupt the sequence of events.
- **Trap:** Incorrect level of detail for diagram elements.
- **Tip:** When deciding the level of detail to include for the description of any diagram element there are several important considerations:
  - *Too little detail——The diagram might be referred to by people separated by time and location from the author (e. g. bowties are often used as a standalone poster). Sufficient detail should be included in the element descriptions so that the reader can generally understand the author's intention without reference to additional explanatory material.*

尽管一些企业确实在企业标准中规定了屏障最低数量，但屏障计数可能适得其反。为满足屏障最低数量的要求，额外的和昂贵的屏障可能被包含进来，而实际并不需要这些屏障；另一方面如果屏障没有有效地控制相关危害，即便是达到了所规定的屏障数量，考虑的屏障也可能太少了。

5.5 识别屏障的陷阱和提示

- **陷阱**：控制措施描述得过于简单，例如：将控制措施仅描述为“安全巡逻”。
- **提示**：描述控制措施具体内容时应时刻考虑到读者，例如：“安全巡逻以识别泄漏并通知控制室”。设法包含中断事件序列所执行的动作。
- **陷阱**：图表元素的详细程度不正确。
- **提示**：为任何图表元素的描述确定要包含的(信息)详细程度时，有几个重要的注意事项：
  - 细节太少——图表可能由与作者在时间和地点上分离的人参考(例如，领结图通常用作独立的海报)。图表元素描述应包含足够的细节，以便读者可以理解作者的意图，而无须参考其他解释性资料。

- *Too much detail——The competing consideration for an appropriate level of detail is that the descriptions should not be overly convoluted or lengthy. Diagram elements may be thought of as risk exclamation marks and by remaining succinct, their communication benefits are maximized. Normally one sentence should suffice.*

- **Trap:** Not including poor quality controls.
- **Tip:** Include controls that are generally considered to be in place even if they have very poor effectiveness. Using colours to depict control effectiveness will highlight these areas for potential improvement.

5.6 References

(1) CGE Risk Management Solutions (www.cgerisk.com).

(2) CGE Risk Academy:

https://www.cgerisk.com/knowledgebase/Main_Page.

- 细节太多——相对于细节太少，图表元素描述不应过于复杂或冗长。图表元素可以是风险感叹号，通过保持描述的简洁性，他们的沟通效益得到最大化。通常，一句话就足够了。

- **陷阱**：不包括有效性差的控制措施。
- **提示**：包括那些笼统来看已实施的控制措施，即便它们的有效性非常差。用颜色描画控制措施的有效性将凸显潜在的亟待提升的领域。

5.6 参考资料

（1）CGE Risk Management Solutions（www. cgerisk. com）.

（2）CGE Risk Academy：

https：//www. cgerisk. com/knowledgebase/Main_Page.

6

LAYER OF PROTECTION ANALYSIS (LOPA)

保护层分析（LOPA）

6.1 Introduction

Layer of Protection Analysis (LOPA) is a semi-quantitative form of risk assessment. It builds on the information developed during a qualitative hazard evaluation, such as a Process Hazard Analysis (PHA) or Hazard and Operability (HAZOP) study. The primary purpose of LOPA is to determine if there are sufficient layers of protection against the consequences of an accident scenario to reduce the likelihood to an acceptable level.

For this, it uses the initiating event frequency and the probability of failure of Independent Protection Layers (IPLs) to estimate the likelihood of a scenario. The frequency of the mitigated consequence can then be compared with the company risk tolerance criteria to determine if the existing IPLs or safeguards are adequate.

The LOPA technique is described in greater detail in the American Institute of Chemical Engineers (AIChE) Center for Chemical Process Safety (CCPS) publication. An overview of the technique is presented here.

LOPA is a recognised technique for determining Safety Integrity Level (SIL) for Safety Instrumented Function (SIF) as required by international standards, i.e. IEC-61508 and IEC-61511.

Safeguard——Features that reduce the frequency of occurrence or mitigate or prevent the consequences.

6.1 简介

保护层分析(LOPA)是一种半定量的风险评估形式。它建立在定性危害评估过程所产生信息的基础上,例如工艺危害分析(PHA)或危险与可操作性(HAZOP)分析。保护层分析(LOPA)的主要目的是确定是否有足够的保护层防备事故情景[1]的后果,是否将其可能性降低到可接受的水平。

为此,它使用初始事件频率和独立保护层(IPLs)的失效概率来估计事故情景的可能性。然后通过将减轻后果的频率与公司的风险承受准则进行比较,确定现有的独立保护层(IPLs)或安全措施是否足够。

保护层分析(LOPA)技术在美国化学工程师协会(AIChE)化工过程安全中心(CCPS)出版物中有更详细的描述。本节呈现了该技术的概貌。

根据国际标准(即 IEC-61508 和 IEC-61511)要求,使用保护层分析是确定安全仪表功能(SIF)的安全完整性等级(SIL)的公认技术。

安全措施——降低事件发生频率以及减轻或预防事件后果的措施。

---

[1] scenario 在本文中称作"情景",在其他文献中也称作"场景"。

IPL——A device, system, or action that is capable of preventing a scenario from proceeding to the undesired consequence regardless of the initiating event or the action of any other layer of protection associated with the scenario. The effectiveness and independence of an IPL must be auditable.

SIF——A combination of sensors, logic solver and final elements that work together to detect an out-of-limit (abnormal) condition and take action to prevent or mitigate the condition.

The relationship between Safeguard and IPL is not always straightforward and can be confusing. In very simple terms, the following can be considered based on the above definitions: A safeguard is anything we have in place to prevent or mitigate something undesirable from happening. Where a safeguard is independent, capable of preventing or mitigating a hazardous scenario and is auditable, it can be considered an IPL.

6.2 Background

An oil and gas processing facility is protected from a hazard through layers of protection or safeguards (Figure 6-1). To be considered a separate layer of protection, the safeguards must be independent of each other [i. e. Independent Protection Layers (IPL)]. The IPLs have certain levels of functionality, availability and reliability. The more hazardous or more complex the process then the more IPLs would generally be required to provide an adequate level of safety.

独立保护层(IPL)——设备、系统或操作，无论初始事件或与该情景关联的任何其他保护层的操作如何，都能够阻止该情景进入不希望的后果。独立保护层的有效性和独立性必须是可审核的。

安全仪表功能(SIF)——传感器、逻辑控制器、最终执行元件的组合，它们协同工作以检测超出限值(异常)的状态并采取措施防止或缓解这种情况。

安全措施和独立保护层(IPL)之间的关系并不总是直截了当，可能令人困惑。简单来说，可以根据上述定义考虑以下内容：安全措施是我们为防止不良事件的发生或减轻不良事件发生的后果所采取的任何措施。如果安全措施是独立的，能够预防或减轻危险情况并且可以审核，则可以将其视为独立保护层(IPL)。

6.2 背景

油气处理设施通过多层保护或安全措施免受危害的影响(图6-1)。要被视为一个独立的保护层，安全措施必须相互独立[即独立保护层(IPL)]。独立保护层(IPLs)具有一定级别的功能性、可用性和可靠性。工艺过程越危险或越复杂，通常需要越多的独立保护层来提供足够的安全水平。

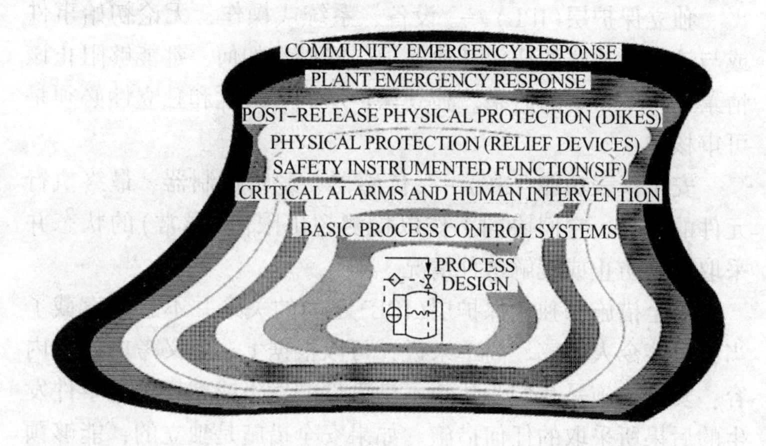

Figure 6-1 Examples of Protective Layers (IEC 61511)

In order to qualify as an IPL, a device, system or action must satisfy the following constraints. It must be:

- Effective in preventing the consequence when it functions as designed.
- Independent of the initiating event and the components of any other IPL which may have already been claimed in the same scenario.
- Auditable——that is, the assumed effectiveness in terms of consequence prevention and the Probability of Dangerous Failure on Demand (PFD) must be capable of validation in some manner.

图 6-1 保护层示例(IEC 61511)

为了有资格作为独立保护层(IPL)，设备、系统或操作必须满足以下约束条件。它必须是：

- 在按设计运行时有效防止后果发生。
- 独立于初始事件和可能已在同一情景中声明的任何其他独立保护层(IPL)的组件。
- 可审核——也就是说，在后果预防和要求时危险失效概率(PFD)方面的假设有效性必须能够以某种方式进行验证。

6.3 The LOPA Methodology

The LOPA methodology calculates the likelihood of a hazardous event using the initiating event frequency and the probability of failure of the installed IPLs. The resultant event likelihood is compared against the company risk tolerance. If there is a gap between the calculated risk level and the tolerable risk then this shows that additional risk reduction is required, i. e. additional IPLs are required. The gap will be in terms of a PFD or a risk reduction factor (RRF), which is the inverse of PFD, which can be specified for the new IPL.

6.4 Uses of LOPA

LOPA:

① Is a method for evaluating the adequacy and effectiveness of IPLs in reducing the frequency and/or consequence severity of hazardous events.

② Provides criteria and restrictions for evaluation of IPLs, reducing the subjectivity of qualitative methods.

③ Provides a consistent basis for determining if protection layers achieve the required risk reduction target.

LOPA is a part of the safety lifecycle activities for IPLs, including Safety Instrumented Systems (SIS). LOPA can also be used to:

① Identify the need for SIS or other protection layers to improve process safety.

6.3 保护层分析(LOPA)方法

保护层分析(LOPA)方法使用初始事件频率和已有独立保护层(IPLs)的失效概率来计算危险事件的可能性。将得到的危险事件可能性与企业风险容忍度进行比较。如果计算得到的风险等级达不到可容忍风险标准，则表明需要额外的风险削减措施，即需要额外的独立保护层(IPL)。差距将以要求时危险失效概率(PFD)或风险削减因子(RRF)表示，后者是前者的倒数，可以为新的独立保护层(IPL)提出条件。

6.4 保护层分析(LOPA)的使用

保护层分析(LOPA)：

① 评估独立保护层(IPLs)在降低危险事件频率和/或危险事件后果严重性方面的充分性和有效性的方法。

② 为评估独立保护层(IPLs)提供标准和限制，降低定性方法的主观性。

③ 为确定保护层是否达到所需的风险降低目标提供一致的基础。

保护层分析(LOPA)是独立保护层(IPLs)安全生命周期活动的一部分，包括安全仪表系统(SIS)。保护层分析(LOPA)也可用于：

① 确定是否需要安全仪表系统(SIS)或其他保护层来改善工艺过程的安全性。

② Determine the target Safety Integrity Level (SIL) for a Safety Instrumented Function (SIF).

③ Evaluate whether protection layers can be considered independent.

If credit is taken during LOPA for an IPL, there shall be associated maintenance, testing, and verification of that layer of protection at the performance level assumed in the LOPA study. The devices forming the IPL shall be recorded in a register of protective devices or HSE Critical Equipment and Systems (HSECES).

Safeguards identified in HAZOP study are the main input into a LOPA study. LOPA should therefore be started either in conjunction with HAZOP study or shortly after issue of the final HAZOP study report. LOPA has the same life cycle as HAZOP study. LOPA studies should be performed in projects and in operations as part of revalidation and Management of Change (MoC).

6.5 Performing a LOPA

There are a number of ways to record LOPA. Software packages exist that provide different functionalities but it is important not to get too engrossed in software. The purpose of the software should be to record the output of the LOPA. This recording can be completed as effectively with pen and paper as with expensive software. It is normal for LOPA to be completed in spreadsheets.

Table 6-1 presents an example of a LOPA table. The information below the table explains each of the inputs (columns) to the table.

② 确定安全仪表功能(SIF)的目标安全完整性等级(SIL)。

③ 评估保护层是否可以被认为是独立的。

如果在保护层分析(LOPA)期间赋予了独立保护层(IPL)“信用数”，则应在保护层分析(LOPA)研究中假定的性能水平上，对该保护层进行相关维护、测试和校核。构成独立保护层(IPL)的设备应记录在保护装置或 HSE 关键设备和系统(HSECES)的登记册中。

危险与可操作性(HAZOP)分析中确定的安全措施是保护层分析(LOPA)研究的主要输入。因此，保护层分析(LOPA)应与危险与可操作性(HAZOP)分析一起启动，或在最终危险与可操作性(HAZOP)分析报告发布后不久启动。保护层分析(LOPA)与危险与可操作性(HAZOP)分析具有相同的生命周期。保护层分析(LOPA)研究应在项目阶段和在设施运行阶段作为重新验证和变更管理(MoC)的一部分进行。

6.5 执行保护层分析(LOPA)

记录保护层分析(LOPA)的方法有很多种。存在提供不同功能的软件包，但重要的是不要过于专注于软件。软件的目的应该是记录保护层分析(LOPA)的输出。与昂贵的软件一样，这种记录可以用笔和纸有效地完成。保护层分析(LOPA)在电子表格中完成是正常的。

表 6-1 列出了保护层分析(LOPA)电子表格的一个例子。表格下方的信息说明了表格的每个输入(列)。

The LOPA starts by recording the risk, severity and likelihood, of the initiating event of the hazardous scenario (columns 1-4 in Table 6-1). It then calculates the likelihood of this event by multiplying it by the probability of failure of the available IPLs (columns 5-8), and by the likelihood of enabling events and conditional modifiers (columns 9-10). Subsequently it compares the resultant risk likelihood (column 11) against the tolerable risk level (column 12) to determine if additional Risk Reduction Measures (RRM) are required.

保护层分析(LOPA)首先记录危险情景初始事件的风险、严重程度和可能性(表6-1中的第1~4列)。然后，它通过将其乘以可用独立保护层(IPL)的失效概率(第5~8列)以及使能事件和条件修正因子(第9~10列)的可能性来计算此事件的可能性。随后，它将得出的风险可能性(第11列)与可容忍的风险等级(第12列)进行比较，以确定是否需要额外的风险削减措施(RRM)。

Table 6-1 Example of LOPA Table

1	2	3	4	5	6	7	8	9	10	11	12	13
Impact Event Description	Severity Level	Initiating Cause Description	Initiation Likelihood (Frequency per Year)	Protection Layers (Probability of Failure)				Conditional Modifiers		Intermediate Event Likelihood	Tolerable Risk Likelihood	Risk Reduction Factor
				Control System	Alarm and Operator Action	Other Protection Devices	Other Mitigation Measures	Occupancy Factor	Probability of Ignition			

表 6-1 保护层分析表格示例

1	2	3	4	5	6	7	8	9	10	11	12	13
影响事件描述	严重程度	初始原因描述	初始事件可能性(频率/年)	保护层(失效概率)				条件修正因子		中间事件可能性	可承受风险可能性	风险削减因子
				控制系统	报警和人员动作	其他保护装置	其他削减措施	人在影响区内的概率	点火概率			

LOPA is based on simplified assumptions regarding the numerical values of each of the components of the hazardous scenario. The risk of the hazardous scenario is approximated using orders of magnitude categories for each of the considered factors (IPLs, modifiers, etc.). The method allows for the identification of the layers of protection that are in place and to identify if additional layers of protection are needed. If additional RRMs are required, the LOPA evaluates the required level of risk reduction (RRF, column 13).

6.6 Advantages and Limitations

LOPA advantages:

① LOPA is effective in assessing layers of protection to manage risk.

② LOPA conforms to international standards IEC 61511, clauses 8 and 9.

③ LOPA facilitates the analysis of protective layers addressing health, safety and environmental risk, and can also be applied to risks due to equipment damage and business value lost.

The limitations of LOPA:

① Unable to identify hazards, analyse escalation events or analyse risks associated with escape and evacuation.

② Excessive for simple or low risk decisions.

③ Overly simplistic for complex systems——where Quantitative Risk Assessment (QRA) and other quantitative techniques are preferred.

保护层分析(LOPA)基于危险情景每个组成部分数值的简化假设。危险情景的风险是用每个考虑因素[独立保护层(IPL)、修正因子等]的数量级类别来近似表达的。该方法允许识别已有保护层，并识别是否需要额外的保护层。如果需要额外的风险削减措施(RRM)，保护层分析(LOPA)将评价所需的风险降低水平[风险消减因子(RRF)，第13栏]。

6.6 优势和局限性

保护层分析(LOPA)的优势：

① 在通过保护层管理风险方面，保护层分析能有效地达成一致意见。

② 保护层分析符合国际标准 IEC 61511，第8和9条。

③ 保护层分析有助于分析处理健康、安全和环境风险的保护层，也可用于因设备损坏和商业价值损失而产生的风险。

保护层分析的局限性：

① 无法识别危害、分析升级事件或分析与逃生和撤离相关的风险。

② 对简单或低风险决策过于复杂。

③ 对复杂系统过于简化——复杂系统首选定量风险评估(QRA)和其他定量技术。

6.7 Conclusion

LOPA is considered, more rigorous, more precise and more resource intensive than other qualitative methods. However, it is seen as less rigorous, less precise and less resource intensive than methods such as a Quantitative Risk Assessment (QRA). The QRA method is not frequently used because it is resource intensive and costly.

It is generally recognised that no single risk assessment technique is adequate for every SIL determination situation, combinations of different methods may be required for a single facility or project. It is becoming more common to utilise robust qualitative or semi-quantitative methods and only utilise the resource intensive, fully quantitative methods, for the high risk, or high consequence scenarios.

6.8 Example of LOPA Table

An example of LOPA table is shown in Table 6-1.

Columns 1-2: Impact Event Description and Severity Level

The accidental event and the potential severity level come from previous hazard identification studies; e.g. HAZOP study, HAZID, etc.

Columns 3-4: Initiating Cause Description and Initiation Likelihood

The initiating cause description of the hazardous event and its likelihood. This will be a single cause of the hazard, not a chain of events. The likelihood is usually expressed as frequency per year.

6.7 结论

与其他定性方法相比,保护层分析(LOPA)被认为更严格、更精确、所需资源更密集。然而,与定量风险评估(QRA)等方法相比,它被认为不那么严格、精确度低、所需资源密集程度低。定量风险评估(QRA)方法不经常使用,因为它所需资源密集且成本高。

通常认为,对于每个需要确定安全完整性等级(SIL)的情况,没有单一的风险评估技术是足够的,单个设施或项目可能需要不同方法的组合。使用稳健的定性或半定量方法变得越来越普遍,而仅仅针对高风险或高后果的情景使用资源密集型、完全定量的方法。

6.8 保护层分析表格示例

保护层分析表格示例见表6-1。

第1~2列:影响事件描述和严重程度

意外事件和潜在的严重程度来自先前的危险识别研究,如危险与可操作性(HAZOP)分析和危险辨识(HAZID)等。

第3~4列:初始原因描述和初始事件可能性

危险事件的起始原因描述以及危险事件的可能性。这将是危险的一个原因,而不是一系列事件。可能性通常表示为每年的(发生)频率。

Columns 5–8: Protection Layer (PL)

These columns include the protection layers to prevent or mitigate the accidental event. An IPL is a device, system or action that is capable of preventing a scenario from proceeding to its undesired consequence independently from the initiating event and any other layers of protection. An IPL must meet these conditions: ① Must be effective in preventing the consequence; ② independent from the initiating event (and other protection layers); and ③ auditable (its effectiveness must be capable of validation in some way).

Columns 9–10: Conditional Modifiers

One of several probabilities included in scenario risk calculations that modify the frequency of the hazardous event having consequences.

Column 11: Intermediate Event Likelihood

The calculated frequency at which the hazardous event frequency would occur with all of the existing IPLs in place (do not include any RRF associated with the scenario being assessed).

Column 12: Tolerable Risk Likelihood

The likelihood corresponding to the specific event severity according to the corporate risk criteria.

Column 13: Risk Reduction Factor (RRF)

This is the RRF to be achieved by the SIF. The RRF relationship to the system's Probability of Dangerous Failure on Demand (PFD) can be calculated as PFDavg = 1/RRF.

第5~8列：保护层(PL)

这些列包括用于防止意外事件或减轻意外事件后果的保护层。独立保护层(IPL)是一种设备、系统或操作，独立于初始事件和任何其他保护层，能够防止危险情景产生不希望的后果。独立保护层(IPL)必须符合以下条件：①必须有效防止后果；②独立于初始事件(和其他保护层)；③可审核(其有效性必须能够以某种方式进行验证)。

第9~10列：条件修正因子

用于修正造成后果的危险事件频率而进行的情景风险计算中包括的几种概率之一。

第11列：中间事件可能性

在所有独立保护层(IPLs)到位的情况下，计算的发生危险事件的频率[不包括任何与评估情景相关的风险削减因子(RRF)]。

第12列：可承受风险可能性

根据企业风险准则，具体事件严重性(发生)的可能性。

第13列：风险削减因子(RRF)

这是通过安全仪表功能(SIF)获得的风险削减因子(RRF)。风险削减因子(RRF)可以用系统的要求时危险失效概率(PFD)平均值的倒数计算得到 PFDavg = 1/RRF。

6.9 References

(1) CCPS, Layer of Protection Analysis-Simplified Process Risk Assessment. American Institute of Chemical Engineers, New York, 2001.

(2) International Electrotechnical Commission (1998). "Functional Safety of Electrical / Electronic / Programmable Electronic Safety-related Systems, Parts 1-7". IEC-61508, IEC, Geneva.

(3) IEC, International Electrotechnical Commission (2001). "Functional Safety Instrumented Systems for the Process Industry Sector, Parts 1 - 3". (Draft in Progress), IEC-61511, IEC, Geneva.

6.9 参考资料

（1）CCPS，Layer of Protection Analysis-Simplified Process Risk Assessment. American Institute of Chemical Engineers，New York，2001.

（2）International Electrotechnical Commission（1998）. "Functional Safety of Electrical / Electronic / Programmable Electronic Safety-related Systems，Parts 1-7". IEC-61508，IEC，Geneva.

（3）IEC，International Electrotechnical Commission（2001）. "Functional Safety Instrumented Systems for the Process Industry Sector，Parts 1-3".（Draft in Progress），IEC-61511，IEC，Geneva.

7

SAFETY CASE

安全例证

7.1 Introduction

A Safety Case provides a documented demonstration that all Major Accident Hazard (MAH) risks have been effectively controlled and that residual risks are reduced to a level which can be considered to be As Low as Reasonably Practicable (ALARP). A Safety Case is a living and evolving record of the design and operation of a facility. The terminology "Safety Case" is often mistakenly referred to as an "HSE Case", however, this is incorrect. While an HSE Case does include the same content as a Safety Case, it has a much wider scope in that it includes both Health and Environmental issues into the document.

A Major Accident is defined as an "Uncontrolled Occurrence" which leads to major or huge catastrophic consequences to people, assets, the environment and/or the company's reputation. The consequences may be immediate or delayed, and may occur outside as well as inside the facility. Major Accident Hazards (MAHs) are those hazards that have the potential to lead to Major Accidents.

7.2 Background and Progression

The origin of the modern Safety Case is the Piper Alpha Disaster (1986) in the offshore UK Continental Shelf. The UK Offshore Safety Case (OSC) Regulations (1992 and 2015) established the regulatory framework for the UK offshore oil and gas industry and heralded a new era moving away from prescriptive rules and standards——based regulation towards a modern risk-based and performance driven approach.

7.1 简介

安全例证提供了一个文件证明，即所有重大事故危害(MAH)风险得到有效控制，剩余风险降低到可以认为合理可行尽可能低的水平(ALARP)。安全例证是与生产设施设计和运行相关的动态的、不断发展的记录。术语“安全例证”经常被错误地称为“HSE例证”，但这是不正确的。虽然HSE例证确实包含与安全例证相同的内容，但它具有更广泛的范围，因为它在文档中包含健康和环境问题。

重大事故被定义为“非受控事件”，会对人员、资产、环境和/或企业声誉造成重大或巨大的灾难性后果。后果可能是立即的或延迟的，可能发生在设施外部和内部。重大事故危害(MAHs)是可能导致重大事故的危害。

7.2 背景和进展

现代安全例证的起源是英国近海大陆架的Piper Alpha灾难(1986)。英国海上安全例证法(OSC，1992年和2015年)为英国海上石油和天然气行业建立了监管框架，并预示着一个新时代的来临——从规范性规则和基于标准的监管转向风险和绩效驱动的现代方法。

The Offshore Installations (Prevention of Fire and Explosion, and Emergency Response) Regulations 1995 (PFEER) that partner with the OSC Regulations were developed to formally support a risk-based and performance (or goal) setting agenda between offshore operators and regulators. The MAH performance improvement successes of the OSC/PFEER approach have led to the widespread and voluntary adoption by the global onshore and offshore oil and gas industry.

The UK regulatory or international Safety Case approach captures a snap shot of MAH performance and can create large volumes of paper (or electronic) documents. These documents often "sit on shelves" and are not referred to during day to day operations. The modern approach is to move away from static Safety Cases towards living and dynamic Safety Cases that are electronic, data driven and based on collaborative action leading to continual improvement. The modern living Safety Case is increasingly focused on real time barrier performance management through direct electronic links to the Computerised Maintenance Management System (CMMS) and the HSE-MS/AIMS.

7.3 Objectives

The objectives of the Safety Case are to:

- Describe the process materials (including raw materials, intermediaries and final products), design, facilities, operations, manning, logistics, meteorological conditions, natural hazards and surrounding environs with MAH potential.

- Describe the facility specific HSE Management System (HSE-MS) and Asset Integrity Management System (AIMS) used to manage MAH risks.

海上设施(火灾爆炸预防和应急响应)法(PFEER，1995年)与海上安全例证(OSC)法一起，正式支持海上油气作业者和监管机构之间基于风险和设定绩效(或目标)的议程。基于OSC/PFEER方法的重大事故危害(MAH)管理绩效的成功提升，引发全球陆上和海上石油天然气行业广泛和自愿采用该方法。

英国法规要求的或国际(普遍采用的)安全例证方法，提取重大事故危害(MAH)管控的大量信息，可创建大量纸质(或电子)文档。这些文档通常"待在架子上"，在日常操作中不会被提及。现代方法是从静态安全例证转向有生命力的动态安全例证，这些例证是电子的、数据驱动的、基于协作行动的、持续改进的。现代的动态安全例证，通过电子方式直接链接到计算机化的维修管理系统(CMMS)和HSE管理体系/资产完整性管理体系(HSE-MS/AIMS)，越来越关注实时的屏障性能管理。

7.3 目标

安全例证的目标如下：

- 描述包含潜在重大事故危害(MAH)的工艺过程的物料(包括原材料、中间产品和最终产品)、设计、设施、操作、配员、物流、气象条件、自然灾害和周围环境。
- 描述用于管理特定资产或设施重大事故危害(MAH)风险的有针对性的HSE管理体系(HSE-MS)和资产完整性管理体系(AIMS)。

- Demonstrate that MAH risks associated with the operation of the facility have been comprehensively identified and that the consequences, threats, escalation factors and barriers required to manage MAHs have been identified and assessed to demonstrate that residual MAH risks have been systematically reduced and are, or will be ALARP. This assessment is generally achieved by producing bowtie models, however, other methods are acceptable.
- Demonstrate that the asset has design, technical and operating integrity and that it is effective. Demonstrate that HSE Critical Equipment and Systems (HSECES) that are in place are effective and available when required.
- Demonstrate that HSE critical activities and tasks have been established and are embedded in the facility processes. Demonstrate that the HSE critical activities and tasks deliver design, technical and operating integrity.
- Demonstrate adequate plans and procedures for onsite emergency response, and offsite crisis management, are in place and that they are based on credible MAH events and scenarios.
- Ensure that omissions and deficiencies in MAH risk management are identified and prioritised in the Remedial Action Plan (RAP). The RAP forms part of the HSE - MS and AIMS continual improvement process.
- Provide a formal "Statement of Fitness" signed off by the Asset Owner that confirms that the management of MAH risks at the facilities is or will be ALARP.

• 证明已全面识别与生产设施运行相关的重大事故危害(MAH)风险；已识别其后果、威胁、升级因素及管理其所需要的屏障；已评估其风险，以证明剩余风险已被系统地降低，是或将是合理可行尽可能低(ALARP)的。完成上述评估一般通过绘制领结模型来实现。但是，其他方法也是可以接受的。

• 证明资产或设施具有设计、技术、操作完整性，并且有效。证明现有的 HSE 关键设备和系统(HSECES)在需要时是有效且可用的。

• 证明 HSE 关键活动和任务已经建立并嵌入到资产或设施的流程中。证明 HSE 关键活动和任务(已经或将会)提供设计、技术和操作完整性。

• 证明适当的现场应急响应计划和程序以及场外危机管理已到位，并且基于可信的重大事故危害事件和情景。

• 确保在补救行动计划(RAP)中确定重大事故危害风险管理中的遗漏和缺陷，并确定补救的优先顺序。补救行动计划(RAP)是 HSE 管理体系(HSE-MS)和资产完整性管理体系(AIMS)持续改进过程的一部分。

• 提供资产或设施运营者签署的正式“适合性声明”，“适合性声明”确认设施中重大事故危害风险的管理是或将是合理可行尽可能低(ALARP)。

7.4 Process

An overview of the Safety Case process is provided in Figure 7-1.

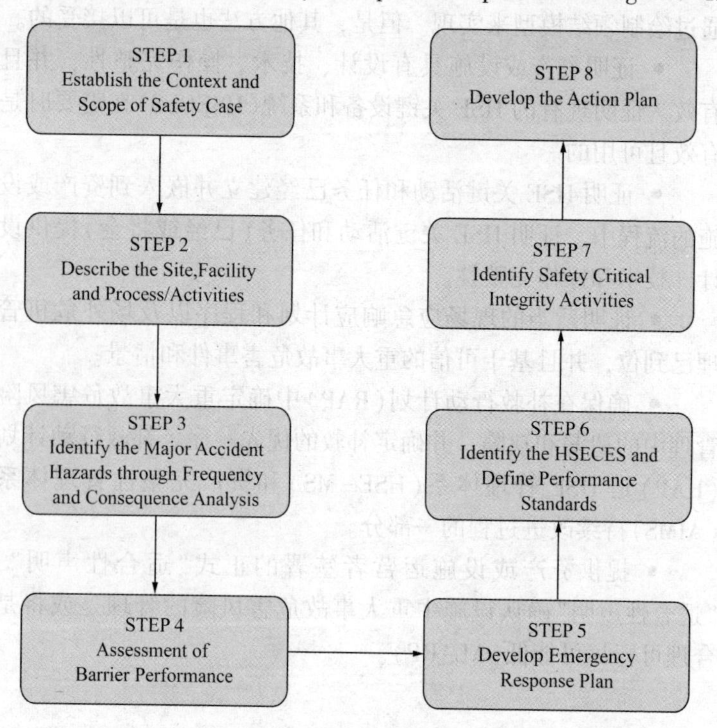

Figure 7-1 Safety Case Process

7.4 过程

图 7-1 中提供了安全例证过程的概述。

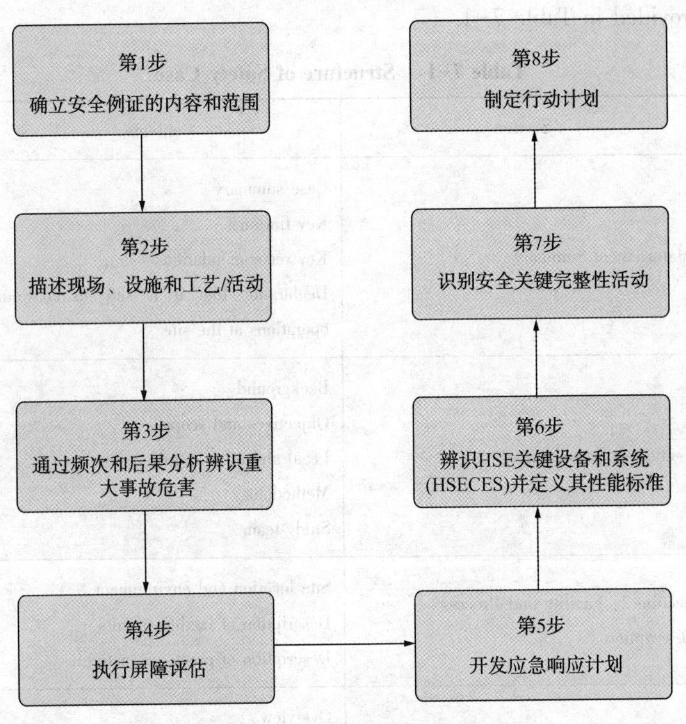

图 7-1 安全例证过程

7.5 Structure and Contents

The typical structure and contents of the Safety Case is provided in Table 7-1.

Table 7-1 Structure of Safety Case

Section	Contents
Management Summary	Case summary Key findings Key recommendations Declaration that it is safe to continue operations at the site
Section 1: Introduction	Background Objectives and scope Legal and other requirements Methodology Study team
Section 2: Facility and Process Description	Site location and environment Description of facilities/units Description of processes/activities
Section 3: Health, Safety and Environment Management System	Overview Safety controls in operations Safety auditing Safety performance review and improvement
Section 4: Review of Hazards and Effects	Overview Safety risk register Identification of major accident hazards

7.5 结构和内容

表 7-1 提供了安全例证的典型结构和内容。

表 7-1 安全例证结构

部分	内容
管理概要	例证摘要 主要发现 主要建议 声明——现场继续操作运行该设施是安全的
第 1 部分：介绍	背景 目标和范围 法律和其他要求 例证方法 编制团队
第 2 部分：设施和工艺描述	设施位置和环境 设施/工艺单元描述 流程/活动描述
第 3 部分：健康安全环境管理体系	概述 操作运行中的安全控制 安全审核 安全绩效评估与改进
第 4 部分：危害及其影响审查	概述 安全风险登记 重大事故危害辨识

continued

Section 5: Bowtie Analysis for Major Accident Hazards	Barrier assessment Bowtie diagrams
Section 6: Safety Critical Elements	Overview Register of SCEs and performance standards
Section 7: Safety Critical Tasks	Overview Register of safety critical tasks
Section 8: Emergency Response	Overview Fire emergency response plan Medical emergency response plan
Section 9: ALARP Demonstration	ALARP criteria ALARP assessment ALARP statement Safety actions and risk reduction Measures Shortfalls and recommendations Remedial action plan
References	List of references used
Appendices	Additional information not included in the main text

7.6 References

(1) UK HSE, The Control of Major Accident Hazards Regulations, 2015.

(2) UK HSE, Offshore Safety Case Regulations, 2005.

(3) UK HSE, Offshore Installations (Prevention of Fire and Explosion, and Emergency Response) Regulations 1995 (PFEER).

续表

第5部分：重大事故危害领结图分析	屏障评估 领结图
第6部分：安全关键部件	概述 安全关键部件(SCEs)[1]登记册及其性能标准
第7部分：安全关键任务	概述 安全关键任务登记册
第8部分：应急响应	概述 消防应急响应计划 医疗应急响应计划
第9部分：ALARP证明	合理可行尽可能低(ALARP)准则 合理可行尽可能低(ALARP)评估 合理可行尽可能低(ALARP)声明 安全行动和风险削减措施 不足之处和建议 补救行动计划
参考资料	参考文献清单
附录	其他未包含在正文中的信息

7.6 参考资料

(1) UK HSE, The Control of Major Accident Hazards Regulations, 2015.

(2) UK HSE, Offshore Safety Case Regulations, 2005.

(3) UK HSE, Offshore Installations (Prevention of Fire and Explosion, and Emergency Response) Regulations 1995 (PFEER).

[1] 安全关键部件与安全关键设备、HSE关键设备和系统在本套丛书中的含义相同。

[illegible]	[illegible]
[illegible]	[illegible]
[illegible]	[illegible]
[illegible]	[illegible]
[illegible]	[illegible]
[illegible]	[illegible]
[illegible]	[illegible]
[illegible]	[illegible]

参考资料

(1) UK HSE, The Control of Major Accident Hazards Regulation, 2015

(2) UK HSE, Offshore Safety Case Regulations, 2005

(3) UK HSE, Offshore Installations (Prevention of Fire and Explosion, and Emergency Response) Regulations, 1995 (PFEER)

[illegible]

APPENDIX

附录

Appendix 1 Leadership Site Visit Prompt Card Sample

LEADER NAME <Insert Leader Name >
SITE NAME <Insert Location>
DATE <Insert Date >

LSV PROMPT CARD
<INSERT TOPIC>

LSV-PC-XX

PROMPT	SCORE
PROMPT <Insert Prompt>	
Supporting Evidence?	
PROMPT <Insert Prompt>	
Supporting Evidence?	
PROMPT <Insert Prompt>	
Supporting Evidence?	
PROMPT <Insert Prompt>	
Supporting Evidence?	

PROMPT	SCORE
PROMPT <Insert Prompt>	
Supporting Evidence?	
PROMPT <Insert Prompt>	
Supporting Evidence?	
PROMPT <Insert Prompt>	
Supporting Evidence?	
PROMPT <Insert Prompt>	
Supporting Evidence?	

附录1 管理人员现场安全督导提示卡样例

管理人员姓名<填写管理人员姓名>

地点<填写地点>

时间<填写年月日>

督导提示卡

<嵌入主题>

LSV-PC-XX

督导提示	打分	督导提示	打分
督导提示 <嵌入督导提示>	打分	督导提示 <嵌入督导提示>	打分
支持性证据?		***支持性证据?***	
督导提示 <嵌入督导提示>	打分	督导提示 <嵌入督导提示>	打分
支持性证据?		***支持性证据?***	
督导提示 <嵌入督导提示>	打分	督导提示 <嵌入督导提示>	打分
支持性证据?		***支持性证据?***	
督导提示 <嵌入督导提示>	打分	督导提示 <嵌入督导提示>	打分
支持性证据?		***支持性证据?***	

SITE NAME

<Insert Location>

LSV PROMPT CARD

<INSERT TOPIC>

LSV-PC-XX

SAFETY CONVERSATION

As part of the site visit you will have safety conversations with the line organisation.
Use this form to summarise your observations.

Who did you talk to? (Provide position not name)
How was your safety conversation received?
How well did they understand the topic?
Were they able to show you procedures they use to control their work?
How well do they understand and implement the procedures?
What information/suggestions did you provide to them?
What improvements did they suggest?

NOTES

LEADER NAME	POSITION
SIGNATURE	DATE

地点

<填写地点>

督导提示卡

<嵌入主题>

LSV-PC-XX

安全交谈

作为现场安全督导工作的一部分，您将与现场人员进行安全交谈。
请使用这张表汇总您的交谈所得。

您与谁进行了交谈？（提供岗位而不是姓名）

你们是如何开展安全交谈的？

他们对本次交谈的主题理解得怎么样？

他们能否向您展示用于控制作业（危害）的程序？

他们对程序理解得怎么样？实施得怎么样？

您向他们提供了什么信息/建议？

他们提出了什么改进建议？

备注

管理人员姓名	岗位
签名	日期

Appendix 2 Abbreviations and Acronyms

附录 2 缩略语

AIChE	American Institute of Chemical Engineers
AIMS	Asset Integrity Management System
ALARP	As Low as Reasonably Practicable
CCPS	Center for Chemical Process Safety
CMMS	Computerised Maintenance Management System
HAZID	Hazard Identification
HAZOP	Hazard and Operability
HSE	Health, Safety and Environment
HSECES	HSE Critical Equipment and System
HSE-MS	Health, Safety and Environment Management System
IADC	International Association of Drilling Contractors
ICI	Imperial Chemical Industries Ltd.
IEC	International Electrotechnical Commission

IOGP	International Association of Oil & Gas Producers
IPA	Institute of Public Affairs
IPL	Independent Protection Layer
ISO	International Standardization Organization
JHA	Job Hazard Analysis
KPI	Key Performance Indicator
LOPA	Layer of Protection Analysis
LSV	Leadership Site Visit
LSVP	Leadership Site Visit Program
MAH	Major Accident Hazard
MoC	Management of Change
NIOSH	National Institute for Occupational Safety and Health
OSC	Offshore Safety Case
OSHA	Occupational Safety and Health Administration
PFD	Probability of Dangerous Failure on Demand
PFEER	Offshore Installations (Prevention of Fire and Explosion, and Emergency Response) Regulations
PHA	Process Hazard Analysis

PL	Protection Layer
PPE	Personal Protective Equipment
PTW	Permit to Work
QRA	Quantitative Risk Assessment
RAM	Risk Assessment Matrix
RAP	Remedial Action Plan
RRF	Risk Reduction Factor
RRM	Risk Reduction Measure
SCA	Safety Critical Activity
SCE	Safety Critical Element
SCT	Safety Critical Task
SIF	Safety Instrumented Function
SIL	Safety Integrity Level
SIS	Safety Instrumented System
SOP	Standard Operating Procedure
TBT	Toolbox Talks
UK HSE	UK Health and Safety Executive
VIP	Very Important Person

Appendix 3 English-Chinese Vocabulary

附录 3 中英文对照词汇表

Action	动作，操作
Actuators	执行机构，执行器
American Institute of Chemical Engineers(AIChE)	美国化学工程师协会
As Low as Reasonably Practicable (ALARP)	合理可行尽可能低
Asset Integrity Management System(AIMS)	资产完整性管理体系
Barrier	屏障
Basic Process Control System	基本过程控制系统
Bowtie Diagram	领结图
Center for Chemical Process Safety (CCPS)	化学过程安全中心
Computerised Maintenance Management System (CMMS)	计算机化的维护管理系统
Conditional Modifier	条件修正因子
Crisis Situation	危机事态
Dangerous/Hazardous Occurrence	危险事件

Detection System	探测系统
Dynamic Situation Hazard	动态情境危害
Emergency Response	紧急响应
Emergency Situation	紧急事态
Enabling Event	使能事件
Escalation Factor	升级因素
Exclusion Zone	禁止区，隔离带
Gripping Device	夹具
Hazard Identification(HAZID)	危害辨识，危险辨识
Hazard and Operability (HAZOP)	危险与可操作性
Health, Safety and Environment Management System(HSE-MS)	健康安全环境管理体系
Health, Safety and Environment (HSE)	健康安全与环境
Hierarchy of Controls	(危害)控制层序
HSE Briefings	HSE 择要说明(会)
HSE Critical Equipment and System(HSECES)	HSE 关键设备和系统
Human Barrier	行为屏障
Ignition Control	点火控制
Imperial Chemical Industries Ltd. (ICI)	帝国化学工业公司

Independent Protection Layer (IPL)	独立保护层
Initiating Event	初始事件
Institute of Public Affairs (IPA)	公共事务协会
International Association of Drilling Contractors(IADC)	国际钻井承包商协会
International Association of Oil & Gas Producers(IOGP)	国际油气生产商协会
International Electrotechnical Commission(IEC)	国际电工技术委员会
International Standardization Organization(ISO)	国际标准化组织
Job Hazard Analysis (JHA)	工作危害分析
Journey Management	旅程管理
Journey Manager	旅程监控人员
Key Performance Indicator (KPI)	关键绩效指标
Layer of Protection Analysis (LOPA)	保护层分析
Leadership Site Visit (LSV)	现场安全督导
Leadership Site Visit Program(LSVP)	管理人员现场安全督导工作，或管理人员现场安全督导方案

Life Saving Equipment	救生设备
Life-Saving Rules	保命法则
Line of Fire	危险轨迹
Loss of Containment	工艺泄漏
Major Accident Hazard (MAH)	重大事故危害
Major Hazard Installation	重大危险装置
Management Audit	管理审核
Management Inspection	管理层检查
Management of Change (MoC)	变更管理
Management Walkthrough	管理巡查
Missed Event	错失事件
National Institute for Occupational Safety and Health(NIOSH)	(美国)国家职业安全卫生研究院
Occupational Safety and Health Administration(OSHA)	(美国)职业安全与卫生管理局
Offshore Safety Case (OSC) Regulations	海上安全例证法
Offshore Installations (Prevention of Fire and Explosion, and Emergency Response) Regulations (PFEER)	海上设施(火灾爆炸预防和应急响应)法
Operational Barrier	操作屏障
Performance Level	性能水平

Permit to Work (PTW)	作业许可
Personal Protective Equipment (PPE)	个体防护装备
Pressure Relief Valve	压力释放阀
Pressurised System	加压系统，承压系统
Prevention through Design	设计保障安全
Preventive Barrier	主动预防型屏障
Probability of Dangerous Failure on Demand (PFD)	要求时危险失效概率
Process Containment	工艺防泄漏
Process Hazard Analysis (PHA)	工艺危害分析
Protection Layer(PL)	保护层
Protection System	保护系统
Quantitative Risk Assessment (QRA)	定量风险评估，定量风险分析
Reactive Barrier	被动反应型屏障
Remedial Action Plan (RAP)	补救行动计划
Risk Assessment Matrix (RAM)	风险评估矩阵
Risk Reduction Factor (RRF)	风险削减因子
Risk Reduction Measure (RRM)	风险削减措施
Safeguard	安全措施
Safety Case	安全例证

Safety Conversation	安全交谈
Safety Critical Activity (SCA)	安全关键活动
Safety Critical Element (SCE)	安全关键部件
Safety Critical Task (SCT)	安全关键任务
Safety Instrumented Function (SIF)	安全仪表功能
Safety Instrumented System (SIS)	安全仪表系统
Safety Integrity Level (SIL)	安全完整性等级
Shutdown System	停车系统
Standard Operating Procedure (SOP)	标准作业程序
Survivability Criteria	耐受标准
Switching Operation	开关作业
Tag Line	牵引绳，溜绳
Toolbox Talks (TBT)	工具箱会议
Top Event	顶上事件
UK Health and Safety Executive(UK HSE)	英国健康安全执行局
Uncontrolled Occurrence	非受控事件
Very Important Person(VIP)	贵宾
Wheel Chock	车轮垫块，轮楔